Perpetual Innovation

PERPETUAL INNOVATION

The New World of
Competition

DON E. KASH

Basic Books, Inc., Publishers

NEW YORK

Library of Congress Cataloging-in-Publication Data

Kash, Don E.
   Perpetual innovation : the new world of competition / Don E. Kash.
     p. cm.
   Includes index.
   ISBN 0-465-05534-6
   1. Technological innovations—Economic aspects—United States.
2. Technological innovations—Economic aspects—Japan.
3. Technology and state—United States.   4. Competition,
International.   I. Title.
HC110.T4K37   1989
338'.064'0973—dc20   89–42513
                                              CIP

To Kelli and Jeff

CONTENTS

PART III

THE JAPANESE ADAPTATION

PART IV

RESPONSE TO THE SYNTHETIC REALITY

PREFACE

THE UNITED STATES is in trouble. While it has established the state of the art in defense, medicine, and agriculture, it is no longer competitive in the international market. How could a country that has won more Nobel Prizes than any other nation, that was first to the moon, that has undertaken the Strategic Defense Initiative, and that invented the microchip, the computer, and the video cassette recorder run a trade deficit in high-tech goods?

How could the United States, with its worldwide reputation for quality goods, have been overtaken by Japan, a country that only thirty years ago was known internationally for its poorly made, imitative products? How could all this have happened so quickly that it caught the United States off guard?

This book looks at the startling transition the United States has experienced in the 1980s and examines why the transition occurred. It investigates the causes of a profound change in the reality of our lives—a change that demands we abandon our traditional ideas about efficiency, the role of the individual, and the utility of the free market.

A key characteristic of the present period is that the speed of technological change has become so great that it is now continuous rather than intermittent. Unlike many of our manufacturing competitors around the world, Americans have not yet adjusted well enough to that fact, either conceptually or organizationally (i.e., by establishing the necessary organizational links among government, industry, and universities). However, making the adjustment is the key to greater economic competitiveness. We need to learn how to be technologically innovative on a continuous basis. This can be accomplished if we develop complexes of organizations that can function competitively within the synthetic reality facing us in commercial markets today. It is a reality where networks of organizations continuously innovate products and processes through synthesis of diverse expertise and materials.

Our basic assumptions about how society operates and about the rules governing success that have served us in the past no longer suffice. Nothing less than a new way of looking at the world—that is, a new vision of reality—is required.

The investigation that follows in this book is divided into four parts:

Part I investigates the nature of the new economic reality we face and the incompatibility of our idea system with that reality.

Part II focuses on how the capability for continuous technological innovation (the basis for the new reality) developed in the United States during World War II and how and why this capability flourished in defense, medicine, and agriculture.

Part III looks at the development of Japan's capability for continuous, commercially oriented innovation and what it means for the economic future of the United States.

Part IV suggests an agenda of actions that the United States must take if it is to prosper in the new reality.

ACKNOWLEDGMENTS

MANY PEOPLE have contributed to this book. I am especially indebted to my students and to numerous colleagues in the Science and Public Policy Program at the University of Oklahoma. Joye A. Swain typed and critiqued more versions than she likes to remember. Lennet Bledsoe, Carol Bernstein, Eileen Hasselwander, and Mary Morrison also provided great clerical and administrative support.

Steve Ballard, Robert Rycroft, Christopher Hill, Vernon Van Dyke, Courtland Lewis, Harvey Brooks, Tom James, Michael Devine, Elisabeth Gunn, Mark Meo, Fujio Niwa, Yoichi Nakamura and Truman Anderson provided critiques of all or parts of the book. The usual disclaimer about their responsibility applies. My thanks to all of the above, and to all the other people who have been willing to talk with me about this project.

PART I

THE SYNTHETIC
REALITY AND
THE SECULAR TRINITY

CHAPTER 1

Contemporary Problems

LIFE today is a paradox. Modern medicine performs miracles: heart transplants are becoming almost routine and cataract surgery is performed on an out-patient basis. At the same time, AIDS threatens a health disaster frequently compared with that posed by the Black Death during the Middle Ages. Rockets carry astronauts into space, but there is no escape from toxic wastes on earth. During the 1980s the American military spent over $2 trillion on sophisticated defense systems, yet the United States is no more able to influence events in the Middle East or Central America now than it was before the money was spent.

During the 1980s the United States has enjoyed the longest-running period of uninterrupted economic growth in its history, yet during that period this nation has moved from being the world's largest creditor to being its largest debtor. Ronald Reagan, a conservative Republican president who was committed to a smaller government, proposed eight budgets with a cumulative deficit of $972 billion, and the national debt doubled. From 1984 to 1987 the value of the American dollar as measured against the currencies of other advanced countries declined by half, but our trade deficit increased. And on October 19, 1987, with unemployment at its lowest level in a decade and with corporate profits strong, the Dow Jones Industrial Average declined by 23 percent.

What lies behind these contradictions? Why have we been unsuccessful in dealing with them?

Ineffective Policy

Ours is a society that confounds the concepts and models that are used traditionally to understand and manage public issues. Indeed, society has become too complex to understand, for everything is hooked to everything else, and whatever actions we take have unexpected consequences. We live in a society where it often seems that the best-informed and most well-intentioned decision makers are as likely to choose policies that will make things worse as they are to choose policies that will make things better.

Perhaps nothing distinguishes the 1980s so clearly as the political system's unwillingness to change policies that handle public problems inadequately. An example is the nation's public education system. Since the 1983 publication of *A Nation at Risk,*[1] numerous studies have documented America's declining relative position in the world. Few Americans would disagree that a poor educational system is costly. It is not just painful to our pride to realize that 20 percent of our population is functionally illiterate while in Japan the figure is 1 percent and in Korea 2 percent, nor is it just a matter for hurt pride that in Japan the average IQ score is 111, while in the United States it is 100.[2] On the contrary, it is nearly universally recognized that high levels of illiteracy are economically and socially costly and that high IQs are economically and socially beneficial. Faced with this widely recognized educational crisis, however, action by the federal government has been minimal in recent years. There has been little public pressure for major new policy initiatives of late, and no major political leader has sought to lead a crusade.

In the 1950s, 1960s, and 1970s, major policy initiatives would have been the standard response to a clearly documented and widely communicated decline of America's educational system. Huge budget deficits no doubt partly explain the lack of public-policy action. But a more compelling reason is a widely held skepticism that public policy can solve public problems. Disillusionment with government's ability to solve problems crested with the election of Ronald Reagan, but it had been building during preceding administrations. In fact, every president since Lyndon Johnson has run against government; financial, airline, and energy deregulation began during the Carter years, for example. It is truly astonishing that many presidents have campaigned against government even while they have headed it.

How is this possible?

Energy illustrates a general pattern. Following the 1973 Arab oil embargo, there was broad public support for energy—and especially oil and gas—price regulation as a way to deal with the economic disorder. Conservatives and those with an economic interest in oil and gas were opposed to regulation for ideological or self-interested reasons, but they were a clear minority. By 1979, when Jimmy Carter began deregulating oil, there was barely a token opposition.[3] Even liberals from consumer states made little more than pro forma statements of opposition. The reason is clear. In the years following the embargo, it became all too evident that government was not capable of effectively carrying out detailed regulation of oil and gas. Indeed, government regulation made things worse rather than better, so that even those who favored regulation for self-interested or ideological reasons became disillusioned. Then a curious thing happened—practical experience overwhelmed self-interest and ideology.

Perhaps an even more striking illustration of the rising concern over the efficacy of government is reflected in the pattern of criticism Barry Goldwater leveled at the Department of Defense during his final period in the Senate, a period when he chaired the Armed Services Committee. Goldwater's continuous and vigorous support of the military was a staple of his Senate career. Yet, in his final days, Goldwater continually expressed concern about mismanagement, inefficiency, and lack of coordination in the Department of Defense.[4] This concern culminated in the Goldwater-Nichols legislation, which increased the power of the chairman of the Joint Chiefs of Staff and of regional military commanders and established a new procurement system for major weapons.[5] Goldwater sought to improve conditions that, from his point of view, threatened the viability of the American military.

Public policy has not lived up to the expectations of most Americans. The ultimate contradiction is that, as we have grown more disillusioned with government, the nation's most compelling problems have increasingly become ones that can be addressed only by the active participation of government. As the nation's public problems become more complex, they require public-policy responses, yet government seems incapable of effectively responding to the changing nature of the problems. The new reality requires new approaches, but to date public policy has not reflected such approaches.

Rhetoric vs. Reality

This book proposes that the nation is faced with a public-policy conundrum. This conundrum is reflected in a discrepancy between the way we believe our society operates and the way it actually operates. A disjuncture has developed between the rhetoric and reality (that is, between the ideology and substance) of our society. The divergence between the ideas and concepts we use to understand and manage society and the reality of the system of production that undergirds society has evolved since World War II. The difference results from basic changes in what and how we produce.

We have experienced a revolution, but it has occurred in an evolutionary fashion and, until recently, its consequences have been so pleasant that we had little reason to be conscious of the revolution. More than two decades ago Michael Harrington articulated what was happening: in previous revolutions the revolutionists "proposed visions which outstripped reality. The unconscious revolutionists of the present create realities which outstrip their vision."[6]

The nation's—and increasingly the world's—productive capabilities are characterized by continuous change that is driven by the capability to carry out continuous technological innovation. These innovations are growing more complex, yet we have very little understanding of the innovation process. Unfortunately, there is a pervasive *illusion* of understanding. Robert Reich of Harvard's John F. Kennedy School of Government has written that, as the pleasant, comfortable revolution evolved, American "ideology was able to accommodate each change and fit it into the old mythic categories with no fundamental revisions in the underlying notions of how America worked."[7] The resulting disjuncture between our idea system and our reality is a significant part of the explanation for our failed public policy.

Successful societies have idea systems with two characteristics: they are shared by the overwhelming majority of the population, and they provide a picture of reality that is reasonably accurate. It is the thesis of this book that contemporary American ideology has the first characteristic but not the second one. The consequence is an increasingly flawed understanding of public problems and decreasingly effective policies. The set of public policy approaches and instruments developed and used from the time of the New Deal until the Reagan administration have been less and less useful as we have experienced a changing reality driven by technological innovation. The Reagan administration's deregulation, free market, and supply-side economic and defense policies appear equally out of touch with the new reality.

Technological Change

Central to addressing the nation's problems and thus to building an effective public policy must be a fuller appreciation of the forces driving change. The changing character of technology and, specifically, technological innovation has become the strongest engine driving society in the 1980s. Three characteristics of contemporary technology and the organizational systems that produce it go far in explaining our most pressing public problems. First, technological innovations regularly generate unanticipated negative consequences. Second, self-conscious technological innovation requires building organizational systems with unique capabilities. Once such organizations have been built, they tend to deliver innovation continuously. Thus innovation becomes a process with its own momentum—i.e., an inertial process. Third, the United States has not built a well-developed organizational system to carry out the continuous innovation of products and processes aimed at winning in the export market. We do not have a system that rapidly converts new ideas and methods into commercial products and processes that are competitive in the world market.

UNANTICIPATED CONSEQUENCES

Technological innovations generate negative unanticipated consequences so regularly that the public has generally come to expect them; indeed, so pervasive is this expectation that the notion of "no free lunch" is part of the currency of debate over technologies. How to deal with both the expectation and the reality of negative unanticipated consequences has become one of the most vexing problems facing the United States.

Two decades ago decisions to develop and utilize technologies and products were made based on their direct and expected impacts. Technologies are normally developed, or at least rationalized, as responses to needs, problems, or markets. Prior to the development of a broad environmental awareness, assessments of the risks associated with new products or technologies occurred for the most part within the confines of specialized technical communities. For example, it has long been recognized that certain vaccines give a small number of people who receive them the disease they are supposed to prevent. Assuming that the numbers faced with such a threat are small, the clear predisposition of the expert community has always been that the broad social benefits far outweigh the limited costs.

One can trace the beginning of a changed perception about unanticipated consequences to the 1962 publication of Rachel Carson's *Silent Spring*, the book widely perceived as having launched the environmental era.[8] What Carson reported on was a growing body of scientific evidence concerning the negative unanticipated consequences of pesticides on the ecosystem. As a result, the public became aware of and sensitive to the phenomenon of unanticipated consequences. The Dalkon Shield represents a different kind of unanticipated consequence. This contraceptive, an intrauterine device, or IUD, was introduced on the market in 1971 and was withdrawn in 1974 when it became clear that the Dalkon Shield was the source of illness—sometimes serious illness—for some of the women who used it.[9] The A. H. Robins Company, producer and marketer of the shield, declared Chapter 11 bankruptcy when numerous lawsuits by those claiming damage from its use were filed. Robins was later taken over by another company.

Contemporary Problems

The number and diversity of products that have triggered unanticipated negative consequences have continued to grow. For both public-policy makers and decision makers within private organizations, new technological initiatives must now be assessed in terms of their potential for unanticipated consequences; but, since the consequences are unanticipated, there can be little confidence in the ability to identify them before the technologies are introduced. Moreover, society is increasingly faced with the need to *correct* the unanticipated consequences of previous actions, the most obvious and pervasive instance being the dumping of toxic wastes. The Congressional Office of Technology Assessment (OTA) has estimated that cleaning up a portion of the 15,000 presently uncontrolled waste sites will likely cost between $10 billion and $40 billion.[10] More troubling, perhaps, is a growing perception that, even if the money were available to carry out the cleanup, there is uncertainty about how to carry it out.[11] One thing is clear: the cost of cleaning up the nation's toxic wastes was not factored into the initial price of the chemicals that produced those wastes. Unanticipated negative consequences, then, frequently require costly and conflict-ridden responses by government and industry. The expectation of negative consequences poses another problem: it acts as a constraint on innovation and, to the extent that rapid innovation is an asset, slowing it down carries social costs. Corporate managers can be forgiven if they do not proceed rapidly with new technologies when they see a previously prosperous pharmaceutical company go into bankruptcy because it introduced a new product.

INERTIAL INNOVATION

The overwhelming majority of U.S. technological innovations occur in defense, medicine, and agriculture.* In these sectors the nation has built large organizational systems that exist for the purpose of continuously producing both new technologies and improvements in existing ones. Together defense, medicine, and agriculture employ well over half the nation's scientists and engineers, spend over 80 percent of federal research and development (R&D) funds (or nearly 90 percent if NASA's

---

*As used in this book, innovation is defined as "the use of new ideas and methods." Thus for innovation to occur, a new product or process must be successfully marketed or deployed.

R&D is included),[12] and accounted in 1985 for 25 percent of the Gross National Product (GNP).[13] In combination, the capacity to innovate continuously and the political support systems that undergird innovation in these sectors pose a fundamental resource allocation problem for the nation. Given the continuous pattern of innovation and the size and entrenched character of the special-interest policy systems in defense, medicine, and agriculture, it is difficult to focus the nation's innovative capabilities on other sectors. This difficulty is inextricably linked to the integration of continuous innovation into our cultural expectations, especially in defense and medicine. In defense, we take perpetual innovation as a given; the "arms race" is merely the label we use for the continuing need to out-innovate the Soviets; the Strategic Defense Initiative (SDI), for example, represents the most recent escalation in the competition for innovative leadership in weapons. This pattern of growth in technologically driven resource demand has characterized the defense sector since the beginning of the Cold War.

A parallel pattern has characterized innovation in the medical sector. Good medical care and technological innovation have become inseparable. Diagnostic and treatment techniques and technologies have steadily increased in number, sophistication, and cost. Similarly, there is always a new "war" to be fought—heart disease, cancer, AIDS, and so on. It is difficult to see a future in which resources from the medical sector will be available for reallocation.

In agriculture new plant species, pesticides, herbicides, and other innovations are continuously available. The result is ever rising production which requires continuous government subsidies for farmers who overproduce.

The nature of the American political system, the expectation of continuous technological innovation, and the investment of most of the nation's innovative capabilities in defense, medicine, and agriculture pose a serious challenge. We appear locked into a future of technological innovation dominated by these three sectors.

DEFICIENT COMMERCIAL INNOVATION

Thus both the problem of unanticipated consequences and the inability to reallocate the resources necessary for technological innovation are closely linked to the third and most urgent problem facing the nation: we have no broad-based, well-developed organizational capability to innovate continuously technologies that are competitive in the international marketplace.

The United States is rapidly losing its capacity to maintain—let alone improve—the well-being of Americans because this nation is not competitive in the world export market. Americans' real spendable weekly earnings declined by roughly 15 percent between 1970 and 1985, while consumption increased. By 1987 we were "borrowing almost 4% of our standard of living from abroad."[14] Two statistics highlight what has been happening to the nation in the world market. Between 1980 and 1987 the nation's merchandise trade deficit grew fourfold, from $36 billion per year to $171 billion per year.[15] To pay for that deficit the United States has tapped foreign financing on a previously unimaginable scale. Between 1982 and 1988 that financing converted the United States from the world's largest creditor to its largest debtor. For example, in 1982 America's net foreign credits were $152 billion, while at the end of 1988 our net foreign debt was $500 billion[16]—greater than the combined debt of Brazil, Mexico, and Argentina, the next three largest debtor nations. In fact, "it is practically inevitable that our net [foreign] debt will reach the $1 trillion mark by the early 1990s no matter how vigorously we act to stem the inflow of foreign savings."[17]

Two factors go far in explaining the rapid erosion in the nation's international economic position. During the last fifteen years the U.S. economy has moved from being relatively self-contained to being the largest component of an increasingly integrated world economy. At the same time, the importance and value of high-technology goods in both the nation's and the world's trade has grown steadily.[18] The growing interdependence of the United States and the world is reflected in the expanded role foreign trade is playing in the nation's economy. Between 1972 and 1987 international trade grew from less than 10 percent of the Gross National Product to 20 percent, and half of corporate profits are now estimated to be derived from foreign trade.[19]

The mix of goods in world trade has been changing. The U.S. product mix is at the leading edge of this process. Between 1981 and

1984 high-tech* goods as a percentage of total U.S. manufacturing trade (exports plus imports) increased from 29 percent to 41 percent,[21] and high-tech goods as a percentage of U.S. manufacturing exports increased from 18 percent to 42 percent between 1980 and 1987. This change in product mix occurred while the nation's export of manufactured goods rose slowly from $160 billion in 1981 to $200 billion in 1987.[22]

During the 1980s, most of the growth in high-tech product exports replaced the loss of market for low-tech products. Unfortunately, while our manufacturing exports were increasing slowly, manufacturing imports were increasing rapidly, with particularly rapid growth in the area of high-tech goods, with imports rising nearly 60 percent during the 1980s. As a result, the United States moved from a high-tech trade surplus of $26.6 billion in 1981 to a deficit of $2.6 billion in 1986. Then, in 1987, the United States enjoyed a small but probably temporary surplus of $0.6 billion, based primarily on increased sales of commercial aircraft.[23]

The erosion of the nation's competitiveness in commercial high-tech products is not so much a result of any decline of capabilities in the United States as it is a reflection of rapidly growing foreign capabilities. Foreign producers of goods for the export market are continuing to develop impressive capabilities for producing innovative new products and processes. The Japanese are currently the recognized leaders in commercial product and process innovation. Indeed, Japan's success has become a model for a number of newly industrializing countries (NICs) that now include Korea, Taiwan, Hong Kong, Singapore, and Brazil.

Although striking in themselves, the data on high-tech exports probably seriously underrepresent the role of technology and technological innovation in America's competitive position. Increasingly, foreign producers are gaining competitive advantage in non-high-tech areas by using innovative high-tech products and processes to produce more and more goods and services. A commonly noted example of the utilization of high-tech products in the manufacture of low-tech goods such as

*The definition of high-tech products used in this book is that frequently referred to as the Department of Commerce–3 definition. High-tech products are those embodying a significantly higher applied R&D expenditure per unit of output than other products.[20]

home appliances is robotics. A more pervasive impact may be seen in agriculture, where technology has contributed to a world-wide food surplus. In the words of Howard Schneiderman, vice-president of Monsanto, "through research agriculture has made the transition from a resource-based to a science-based industry."[24]

A review of the nation's trade performance in sixteen product categories in the six years from 1980 to 1985 shows how badly our competitiveness has eroded (see Appendix, figures 1 through 16). It is useful to set this erosion in context by noting the pattern of development of technological innovation in other countries. Doing so will help explain how the competitiveness problem became so serious for the United States and why it is likely to remain serious. First, America's major commercial competitors give exports primary focus in their national policies. In Germany, Italy, Sweden, Japan, and the NICs, as well as in other countries, competitiveness in the export market is a major— and in some countries the highest—national policy objective. In fact, exports have been the keystone of Japan's national policy since the end of World War II. Thus in the early 1950s the Japanese government formally identified the survival of the nation with "its ability to develop exports to cope with the lack of raw materials and the chronic balance of payments deficits."[25]

The Japanese development pattern is of particular importance because in it are the seeds of an even more intensely competitive future for the United States. As Japan increasingly perceives threats to its export position from the low-labor-cost NICs who are following its model, the Japanese are striving to accelerate their rate of technological innovations. If foreign commercial competition in the high-tech area has been unsettling for the United States to date, the future will likely be much more disturbing. For that reason, regaining a competitive position in world trade is this nation's most compelling challenge.

Synthesis

If the United States is to deal effectively with the problems it faces in the late 1980s, it must recognize clearly how the process of technological innovation has changed. What makes the contemporary period unique is the ability in a growing number of countries to carry out technological innovation on a continuing basis. This capacity for innovation rests on a capability to synthesize products and processes. That is, several countries now have organizations capable of combining materials, ideas, knowledge, and skills continuously in ways they have never previously been combined to produce products and processes that never existed before.

The United States excels in this capability in defense, medicine, and agriculture, but if we are to address the nation's current problems, we must begin to develop our synthetic capabilities in other sectors. What has evolved in the United States in defense, medicine, and agriculture and in a number of other countries in the area of commercially oriented innovations are production systems that work differently from those of the past.

Our obsolete idea system is a serious barrier to an effective national response to trade erosion. The United States will be competitive only when national policy stimulates technological innovation aimed specifically at the export market.

It is not difficult to understand how our reigning ideology became obsolete. In the three decades following World War II, the United States dominated commercial innovation using spinoffs of defense innovations which were developed to serve the domestic market. This pattern of domestically focused commercial spinoffs occurred in a context of consensus among policy makers that no special policies or programs aimed at inducing commercial innovation were necessary.[26] The base of technology and the organizational complex generated by the need for defense innovation, plus the inherent drives of the free market, combined to render the United States dominant in the international high-tech market. However, as figures 1 to 16 (see Appendix) suggest, the patterns of the past can no longer support U.S. leadership.

Contemporary Problems

The nation's deteriorating trade balance and the recent precipitous change in trade in high-tech goods underscore the seriousness of the problem.

The continued dominance of an obsolete ideology has three increasingly evident costs. First, we are precluded from formulating the deliberate policies necessary to make ourselves competitive in the world. Those policies must recognize that only government/private-sector cooperation and support can lead to competitiveness. Second, the pattern of government/private sector cooperation in defense offers important lessons in how to formulate a response to the trade problem, but we are not using that model. Third, although defense innovation is a continuing success story, the growing efforts to force it into the categories dictated by our obsolete ideology threaten that success. This point will be investigated in Part II.

CHAPTER 2

Innovation: The New Competitive Environment

FEW WILL DISAGREE that we live in a world whose people have a growing capacity to synthesize new or improved products, processes, and projects continuously. Those who have lived through the post–World War II period have witnessed a continuous stream of new consumer electronics devices, aircraft, drugs, and higher-yield plant species, to name but a few. Americans now take continuous synthetic innovation, or at least its fruits, as a given; we expect technology to deliver an infinite stream of new products, and we assume that technology can solve many of our problems. Harvard economist John Kenneth Galbraith expressed this characteristic of the contemporary period as follows: "It is a commonplace of modern technology that there is a high measure of certainty that problems have solutions before there is knowledge of how they are to be solved."[1] Clearly, the nation's economic problems do not stem from a failure to recognize the importance of technological innovation. Rather, they have resulted from a failure to appreciate the changed way in which synthetic goods are created and produced. The starting point of a search for answers is a greater appreciation of the difference between the industrial society and the synthetic society.

The industrial society was organized around and structured to manage a distinctive way of producing goods. As described by Robert Reich, it was a society "geared to manufacturing large quantities of relatively simple, standardized products. The key was long production runs with each step along the way made routine and simple. As the scale of production increased, the cost of producing each unit plummeted."[2] Industrial societies are characterized by divisions and subdivisions: success comes from dividing known wholes into parts and parts into components. The goal of these societies is efficiency, achieved by optimizing the production of each component to make the whole. With the production of each component optimized, the production of the whole is optimized.

By contrast, the synthetic society is organized so as to combine components into previously nonexistent wholes, be they products, processes, or projects. The goal of the synthetic society is innovation. That is, its goal is to create newer, higher-performance products or to achieve higher quality in existing products more cheaply by integrating components in the production process in new ways.

Origins

The origins of the synthetic society are, in part, traceable to the capacity to manipulate atoms and molecules.[3] When atomic physicists began to break new ground in the early twentieth century, a strikingly different view of the nature of matter developed. Using new experimental equipment, physicists evolved theories that provided remarkable insight into the properties of subatomic particles and the ways they interact to generate the characteristics of the atom. Using their new understanding and developing ever more powerful experimental equipment, atomic physicists released powerful energetic particles from the nucleus, thus unleashing both artificial and natural radioactivity. By the time of World War II, atomic physicists had theorized that vast quantities of energy could be released from the nucleus of the atom, and it was on this theoretical basis that nuclear weapons and nuclear reactors were ultimately built. These new theories also led to deductions about

17

how atoms would behave in solids. Based on these deductions, experiments were carried out, and out of this process came the proliferating solid-state electronics capabilities we know today.

Although its origins were earlier, a similar capacity for creativity was being demonstrated in organic chemistry. For example, organic compounds had been known as early as the eighteenth century; then, in 1824, urea became the first organic compound to be synthesized. As knowledge and experience accumulated, chemists began to synthesize an ever larger number of organic molecules so that, by the mid-nineteenth century, this process of synthesis had led to the aniline dye industry in Germany.[4] Chemists had begun by creating organic molecules in order to learn more; before long, they were creating new molecules to serve particular purposes, and a broad-based chemical industry was launched. Biologist Barry Commoner of Washington University characterizes what occurred: "Thus the pre-war scientific revolution produced in modern physics and chemistry sciences capable of manipulating nature—of creating for the first time on earth wholly new forms of matter."[5]

During World War II what had started as an ability to synthesize new materials in chemistry and physics exploded into a broad capacity in the United States to synthesize new products, processes, and projects in general. By the 1960s, we had moved to a point where not only advanced computers but even programs like Apollo could be synthesized. For the first two decades following World War II, the development of a broad-based capability for synthesis grew in the United States with little foreign competition. Then, over the course of the post–World War II years, the route to commercial success changed. By the 1970s, other countries, led by Japan, had developed their own synthetic capabilities aimed specifically at innovating products and processes that could compete in the international marketplace, and the capacity for synthetic innovation had taken on major importance. The continuous innovation of new or superior commercial products and processes had increasingly become the basis for competitive leadership.

Characteristics of the Synthetic Reality

The growing capability to synthesize products, processes, and projects has created a contemporary reality with four distinctive characteristics. First, ours is a society pervaded by synthetic, i.e., man-made, products—plastics, synthetic fabrics, synthetic foods, and synthetic information systems are ubiquitous and relatively inexpensive. Natural products, on the other hand, seem to have become status symbols for the well-to-do; therefore, those who wish to eat natural foods and wear natural fibers must usually pay a premium. So commonplace are synthetic products that many of the young in our society regard natural products as fads, as items to be desired simply because they are different. In the immediate post–World War II period, most synthetic products were American-made. Today a large and growing number of them are made elsewhere.

The second characteristic of the synthetic society is the ability to produce synthetic substitutes not only for natural products but for existing *synthetic* products. For example, faced with the lack of natural rubber during World War II, the United States developed synthetic rubber made from oil. Interestingly, rubber can be made from soybeans or even from trash—from any organic material, or even some inorganics.

The capacity to synthesize substitutes has had striking consequences. For example, it has altered the age-old concern about the scarcity of natural resources. Faced with shortage, contemporary society synthesizes a substitute. Malthus has been stood on his head. Perhaps the most striking illustration is a world awash in food. Not only is massive starvation in India no longer an immediate threat, India was ironically a net exporter of certain grains in 1986. Furthermore, the capacity to synthesize threatens all existing products and processes with obsolescence. Both natural and existing synthetic products can be made obsolete by synthetic innovations that produce either higher-quality or higher-performance products, frequently at significantly lower cost. For instance, the transistor made the vacuum tube obsolete, then the silicon chip made the transistor obsolete. Thus the capacity to synthesize at once freed society from the threat of scarcity and threatened it with rapid obsolescence. No product is now secure from the threat that competitors will produce it more cheaply or produce a better alternative. Increasingly, the source of that threat for American products is coming from overseas as foreign nations have developed their own capacity for synthetic innovation.

Third, in those sectors, whether in this country or overseas, where organizational complexes with synthetic capabilities exist, new or superior synthetic products and processes are produced continuously; technological innovations are delivered steadily like an auto assembly line delivers cars. Indeed, since World War II the defense sector has delivered generation after generation of new weapons systems. More recently the Japanese have delivered generation after generation of consumer electronics products. Preexisting demand is no longer considered a requirement; rather, products and processes have become the generators of their own demand.

Fourth, the capacity to synthesize offers a totally new capability—the capability to create on demand. That is, the synthetic capability permits society to do things that have never been done before with the confidence that they can be accomplished within prescribed periods of time. We have witnessed the ability of the United States to put a man on the moon at a time when it did not know how to do it and to accomplish that task within a predetermined period. The Strategic Defense Initiative (SDI) represents only the latest large-scale illustration of the perception that a capability exists to create on demand. It has been suggested that optimists hope we can innovate SDI, while pessimists know we can. Such undertakings, however, demand huge resources, precluding the use of those resources for other purposes.

The new synthetic reality is dominated by organizational complexes that have as their reason for being continuous synthetic innovation—specifically, the innovation of complex technologies that are more dependent on knowledge than on materials. If we are to adjust to and effectively manage in the future, we must recognize that technological innovation is an organizational product, and so, in order for continuous innovation to occur, organizational complexes with unique capabilities must be built. Where such organizational complexes have been built, synthetic innovation occurs continuously. In their absence, innovation is rare if not absent.

A growing number of foreign competitors, led by the Japanese, are organized to carry out continuous synthetic innovation of consumer products. The lack of a similar capability in the United States is a major contributor to our declining competitiveness in the world market, as described in chapter 1. Clearly, basic changes are necessary if we are to escape a growing economic malaise.

The barriers to change derive from a combination of ideology and inertia. An ideology that emphasizes the free market, individualism, and efficiency prevents us from building the cooperation among industry, government, and universities that is required to synthesize commercial products. The inertia results from the fact that a large portion of our scientific-technical resources are committed to the organizational complexes that carry out synthetic innovation in defense and medicine; therefore, they are not available for commercial pursuits. These barriers must be overcome if we are to organize on an adequate scale for synthetic innovation in the commercial export sector. Only such a capability will make us competitive in the vital export race.

The U.S. Capability

The United States led the way to the synthetic reality by developing organizational complexes with capabilities for continuous technological innovation in defense, medicine, and agriculture. Three characteristics distinguish the powerful organizational complexes that carry out synthetic innovation in the United States. First, they tie together government organizations, industrial organizations, and universities. These complexes are distinguished by their ability to tap and integrate needed knowledge, skills, and capabilities rapidly wherever they exist and, when they do not exist, to create them. Second, synthetic innovation requires the integration, i.e., the synthesis, of knowledge from a diverse mix of disciplines, of skills from a diverse mix of experience, and of materials never previously combined. Synthesis involves more than the ability to tap specialized knowledge and skills to produce innovative products and processes; it requires the generation of synergism. Synergism involves combining expertise so that the output represents more than the sum of the specialized knowledge and skills. Synthesis occurs within organizational complexes that utilize interlinked sets of groups, teams, committees, and task forces to produce innovation. Thus in the 1980s innovation has seldom been the product of an individual's creative genius; it has increasingly become the product of groups. In truth, many members of the complexes are unaware that they play a role in innovation, even though it may be a key one. A good example is technical-standards–setting bodies. Third, these complexes depend on government's absorbing a major portion of the high-risk front-end costs—the R&D—and providing an assured market for the innovative product.

Innovation in defense, medicine, and agriculture is achieved by groups working within systems characterized by a nearly seamless web of organizational interactions among government, industry, and universities. In defense government has paid for the continuous conceptualization, design, and development of new technologies with huge R&D expenditures and then has provided a market for the mature technologies. For example, not only did government pay for much of the work leading to the development of the silicon microchip but NASA also provided the large market that allowed Fairchild and Texas Instruments to learn how to mass produce high-quality, low-cost chips.[6] A more recent example of the same pattern is gallium arsenide circuitry. In medicine technological innovation has rested heavily on the large R&D program funded by the National Institutes of Health, but government has also assured a market for technologies through its health insurance and hospital construction programs. Any new medical technology or technique has an automatic market once it has been approved by the medical community, and cost is not a factor in the approval.

Finally, in agriculture government has supported R&D continuously for nearly a century and has, for the last fifty years, stimulated innovations through farm programs designed to manage production (supply). Through this management of production, government has insured that agricultural innovation remains profitable.

Evolution in Manufacturing

The industrial-to-synthetic evolution is most clearly reflected in the operation of worldwide manufacturing organizations and in the products they produce. Manufacturing can no longer be dealt with usefully as a single activity—it must now be thought of as a continuum, a procession of organizations, products, and processes extending from those rooted in the Industrial Revolution to those that are part of the more recent trend toward continuous synthetic innovation. Although classification is necessarily arbitrary, it can be the first step toward clarity. Thus it is useful to divide manufacturing into two different systems: an industrial system and a synthetic system. Most steel making is part of the industrial system, for example, while electronics is part of the synthetic system. Today the synthetic system is growing and dynamic; the industrial system is declining.

Two developments explain the accelerating importance of the synthetic system. First, it is the creator and producer of a continuously advancing, ever-widening stream of new or superior products and processes. Second, as a growing number of new synthetic products and processes are integrated into the manufacture of traditional industrial products, traditional products are becoming part of the synthetic system. Automobiles represent the evolution from the industrial system to the synthetic system, since plastics, composites, ceramics, and solid-state electronics have been integrated into them just as robots and just-in-time inventory systems have been used in their manufacture. The relative decline of those manufacturing sectors that cannot make the transition into the synthetic system seems certain.

Synthetic products and the organizational complexes that produce them—the synthetic system—are increasingly the dynamic component of our society and economy. The media refer to the two systems as "smokestack industries" and "high-tech industries." Any contemporary college graduate understands which one offers real opportunity— smokestack industries are declining; high-tech industries are growing. Smokestack industries were the engines driving the industrial society; high-tech industries are the engines driving the synthetic society. Both domestically and internationally, the ability to carry out synthetic innovations has become steadily more important since the end of World War II.

The capacity for continuous synthetic innovation has changed the nature of international competition. More and more, the pattern of innovation reflected in products aimed at the export market is characterized by rapid rates of turnover, with each turnover representing a substantive difference in products and/or processes. In this pattern, the international commercial manufacturing sector has taken on the characteristics of defense, medicine, and agriculture in the United States. Innovation has made change a permanent condition.

Innovation

Innovation requires novel ideas or methods *and use.*[7] Novel ideas or methods may be reflected in one, a combination, or all the steps in a sequence that includes conceptualization, design, development, production, and use. Innovation has one or some of the following objectives: (1) lower cost, (2) higher quality, (3) superior performance, (4) additional performance, or (5) new performance. Whatever else may be involved, innovation only occurs when ideas are used; i.e., it must include deployment and/or marketing. Thus it involves more than conceptualization, design, and development, i.e., the creation of prototypes. Innovation is different from invention, although it normally requires invention. Further, the innovator may be different from the inventor, as the Japanese have demonstrated time after time, as in the case of the videocassette recorder. In the export market, competitive advantage goes to those who are first or early in production and marketing, not to those who invent, just as in the arms race competitive advantage goes to those who deploy new weapons systems first.

Innovation is not a new phenomenon, nor is it unique to the synthetic system. Quite to the contrary, the industrial revolution was the result of innovations beginning with the steam engine and progressing through such advances as the automobile and the light bulb. What distinguishes industrial and synthetic systems in terms of innovation is (1) the extent to which innovation is planned; (2) the source of innovation; (3) the source of competitive advantage; (4) the nature of products and processes; and (5) the rate at which innovation occurs.

Innovation in the industrial system is not planned; rather, it is random. Both the innovations that brought about the industrial revolution and their socioeconomic impacts came as surprises, as they were unplanned and unexpected. In contrast, in the synthetic system, innovations are planned and routine. They are not surprises; indeed, they are anticipated. In the industrial system, too, innovation is generally the product of individual geniuses, early examples of whom were Watt, Ford, and Edison. In the synthetic system, innovations are rarely the creation of individuals. Rather, they are delivered by groups interacting in organizational complexes. Competitive advantage in the industrial system flows from the pursuit of efficiency. Thus when competition exists, advantage goes to the producer with the lowest costs. In the synthetic system, however, competitive advantage flows from the pursuit of innovation rather than efficiency. Those first or early with the new or superior product/process have a competitive advantage.

The nature of the products manufactured by the industrial and synthetic systems is different. The industrial system manufactures relatively simple, standardized products whose basic character remains the same for long periods, and it uses relatively unchanging processes to produce its goods. The synthetic system produces a continuing stream of either new or qualitatively different products that have superior or new performance capabilities, or else it manufactures old products using unique processes designed to deliver higher quality and lower cost.

Perhaps the most important distinction between the industrial and synthetic systems is the rate at which innovation occurs. In the industrial system, innovations, once accomplished, remain relatively unchanged for long periods of time. Initial product and process innovations normally become the focus of improvements aimed at increased efficiency, but their basic character remains unchanged. The steam engine and the light bulb are essentially the same today as when they were first mass produced. The innovations that led to the industrial revolution made old products and old ways of doing things obsolete.

For instance, the automobile made the horse obsolete, and the light bulb made the kerosene lamp obsolete. Then the assembly line made single-unit manufacture obsolete. An innovation-to-obsolescence cycle occurred though it was a cycle that took decades to complete. Innovation is continuous in the synthetic system. The marketing in 1987 of digital audio tape players came only three years after compact discs revolutionized sound systems.[8] This high rate of change was no surprise in the rapidly changing consumer electronics sector. Moreover, a pattern of rapid and continuous innovation now characterizes a large and growing number of product areas. The process of continuous innovation that characterizes the synthetic system is rapidly shortening the life span of innovation-to-obsolescence cycles—from a few months for software packages to ten to fifteen years for fighter aircraft.

Organizations and products with the characteristics exemplifying the synthetic system are beginning to dominate the world. Since competitive advantage goes to those who innovate most rapidly, products that become the focus of synthetic innovation tend to experience accelerating innovation-to-obsolescence cycles.

Radical vs. Incremental Innovation

To appreciate the nature of competition in the synthetic society, it is necessary to distinguish between two categories of innovation: radical and incremental. Radical innovations allow tasks to be accomplished in ways that were previously not available, or they make it possible for totally new tasks to be accomplished. For example, the intercontinental ballistic missile (ICBM) provided a new way to deliver strategic weapons, whereas previously bombers were the only option. If fully successful, the Strategic Defense Initiative would offer for the first time the ability to neutralize the Soviet ICBM threat. In the commercial realm, recent investor interest in biotechnology is driven by previous examples of radical innovations such as Xerox copiers. Successful radical innovation offers great competitive advantage and high payoff, but it is rare.

Incremental innovation is the more common form of innovation. It involves the modification or redesign of a product or process so as to result in lower cost, higher quality, superior performance, additional performance capabilities, or some combination of these. Commonly incremental innovations result from use of a superior subsystem or component or an additional subsystem or component. Examples range from the substitution of the jet engine for the propeller engine to the addition of air-to-surface missiles on fighters to new circuit boards in Walkmans. In the case of the latter, according to *Business Week*, Sony designed "a new model that cut more than 30 percent off production costs partly by integrating the playing mechanism onto the printed circuit board. The basic $32 Walkman has become Sony's best selling model in the U.S."[9] In the Walkman case innovation involved the combination of previously separate elements into a simpler system with fewer elements.

In both defense and the high-tech export market, radical innovations have commonly been the launching pads for continuous incremental innovation. The RAM chip has evolved through a sequence that has included 1K, 4K, 16K, 64K, and 256K chips with the 1-megabyte chip now coming on line. Expectations are for 4- and 16-megabyte chips to follow. At the other end of the spectrum, incremental innovations are beginning to revolutionize many traditional products. Synthetic innovation is becoming a source of substitutes for traditional materials and components being used in established product lines. The most evident examples are the growing number of substitutes for steel. Here the rate of turnover is slower than for high-tech products, but there is evidence of acceleration. One need only look at the changing mix of materials in automobiles to gain a sense of what is occurring in manufacturing in general. As flexibility in both the production and use of new materials increases—that is, as manufacturing processes and techniques are innovated, making it easier to integrate new components and materials—the rate of substitution will likely increase.

The Competitive Cycle in a Synthetic Society

Once a product or technology area in the synthetic society becomes the focus of an organizational network that has the capacity for synthetic innovation, the pattern is for incremental innovation to follow incremental innovation. Each incremental innovation represents both the termination of the preceding cycle and the beginning of the next one. Each successive innovation serves to reduce demand for the previous product rapidly and to create demand for the new product. This phenomenon is referred to as "technological obsolescence." It is the superior or additional performance capabilities of the incremental innovation that both reduce the old demand and create the new demand. Computers represent a classic illustration. Anyone who has a five-year-old business or personal computer has an obsolete model—not because it works less well or does less than when it was purchased, but because the newer models process more data more rapidly and/or have additional capabilities, such as the rapid generation of high-resolution graphics.

As the technological obsolescence phenomenon has been repeatedly demonstrated, buyers have made innovation-to-obsolescence cycles their framework for decision making. Superior or additional performance capabilities are expected to provide additional benefits to buyers who use the product to perform tasks; therefore, buyers have powerful incentives to buy nothing less than state-of-the-art technology. Even when they see no immediate need for such up-to-the-minute capabilities, they are hesitant to buy less since they assume that in a rapidly changing world they will need the capabilities in the future. Finally, it must be noted that in the synthetic society state-of-the-art technology has taken on status significance, and buying less than that is like buying clothes that are out of fashion.

Given continuous innovation, and thus continuous technological obsolescence, the innovation-to-obsolescence cycle becomes the framework of competition. The organizational complexes that carry out synthetic innovation understand that the iron law of competitive success is to be first or early in deployment or marketing. In the commercial high-tech market, success flows from an ability to create and satisfy demands that develop rapidly and disappear rapidly.[10] The benefits are multiplied if the innovation-to-obsolescence cycles are reinforced by products that have superior or new performance capabilities *and* lower cost with each incremental innovation. Such has been the case in consumer electronics. It must be emphasized, however, that lower cost is not necessarily the critical factor. In many high-tech areas, for example, scientific instruments, costs have risen with each incremental innovation. The critical competitive element is performance, with cost, quality, and the ability to carry out tasks all being viewed as performance measures. Innovation, then, offers two competitive benefits: it creates its own market, and it reduces the competitor's market if he/she does not innovate in a parallel time frame.

In defense, the advantage of being first in deploying radical innovations and fast in deploying incremental innovations is that speed provides power advantages. For example, it is widely believed that a key reason the Soviets backed down during the Cuban missile crisis was the existence of a significant number of operational missiles in the United States at a time when the USSR had only four.[11] Currently, the politics of SDI appear to reflect the same dynamic.

The Rise of Innovation

The emergence of synthetic innovation as the critical instrument of competition in the years following 1945 seems to have taken place in three cumulative phases. Each phase is linked to and rests on the preceding one. The phases are distinguished by two factors: the nature of the key competitors and the substantive focus of the innovation.

Phase One began with the Cold War in the late 1940s and involved a competition between the United States and the USSR for innovative leadership in military technology and weapons systems—a competition that continues unabated today. Phase Two began in the 1950s and involved competition among American companies for innovative leadership in commercial products that were in large part spinoffs from the arms race. Phase Three began in the 1970s and involved competition between American and foreign (particularly Japanese) companies for leadership in the incremental innovation of commercial products many of which were initially produced exclusively in the United States. During the 1980s a growing number of countries have entered this competition, and more will likely enter in the 1990s.

Beginning in the 1970s the United States faced a second major international competition. An export race had been added to the arms race. Both competitions have their origin in the Cold War. In both, success or failure rests on the ability to carry out rapid and continuous synthetic innovation. The arms race, with its associated commercial spinoffs, has been both the resource base and the training ground for the newer export race.

THE ARMS RACE

The arms race had its roots in the development of new military technologies during World War II and in the expectations they created. Clearly the most significant development was the atomic bomb, an innovation representing a quantum leap in explosive capability. The combination of one bomb and one airplane resulted in damage equivalent to what had previously required continuous saturation bombing to accomplish. This revolution resulted from a weapon that had not even been imagined by the military when the war began.

The bomb was the product of an organizational complex that was built deliberately to synthesize it. This secret innovative effort, carried out under the code name Manhattan Project, changed expectations fundamentally and for all time. With its use on 7 August 1945 the atomic bomb made all previous thinking about warfare obsolete. In his powerful book *The Wizards of Armageddon*, Fred Kaplan summarizes the impact as follows: "The whole conception of modern warfare, the nature of international relations, the question of world order, the function of weaponry, had to be thought through again."[12] It was now realized that technologies not available at the beginning of a war, and perhaps not even thought of, could determine its outcome. Thus in all subsequent conflicts, whether "hot" or "cold," political and military leaders would have to plan for continuous innovation. A new specter hung over mankind: technological innovation could fundamentally change the balance of power. New technologies might make the weak strong and the strong weak, and they might do it over short periods of time. Anyone who doubts this has only to ask what the outcome of World War II would have been had our enemies developed and used the first atomic bomb. The rapid innovation-to-obsolescence cycle had become a permanent expectation.

The new perception of the world, conceived by military strategists over the course of the Cold War, was one in which power flowed from innovation. The lack of a capacity to innovate continuously threatened a nation with military obsolescence. This new perception was to become the dominant ingredient of the Cold War arms race. In order to understand the adoption of innovation as the prime instrument in national defense policy, it is useful to take a closer look at the evolution of the arms race. The Cold War began with a perceived Soviet military threat to Western Europe. With a small occupation force located in Germany, the United States and its European allies found themselves facing overwhelming numbers of Soviet and East European forces. To counter the Soviet threat with conventional forces would have required mobilizing and stationing numerous troops in Europe, with the recognition that they would have to be maintained for the indefinite future. That option was both politically and economically unacceptable. Thus the option chosen to counter the Soviet threat was superior military technologies—specifically, the United States's atomic monopoly and its superior strategic bomber force. The first U.S. technological advantage was threatened when, within a year of the Berlin blockade, the Soviets successfully tested an atomic bomb and the United States lost its nuclear monopoly. In response to the new Soviet atomic capability, the United States initiated development of the hydrogen bomb, a bomb with superior performance capabilities. With the successful testing of the hydrogen bomb, an innovation-to-obsolescence cycle occurred, and the atomic bomb became obsolete. The Soviets responded in turn with a program to develop their own hydrogen weapons.

An even more impressive illustration of these continuous innovation-to-obsolescence cycles is represented by the evolution of strategic delivery vehicles.[13] At the beginning of the Cold War the U.S. strategic delivery capability was a bomber force consisting of the B-29 and the

B-36. Both incremental and radical innovations followed as bombers changed incrementally and entirely new delivery vehicles were developed: during the 1950s the B-29 and B-36 bombers were made obsolete by the development of the B-52; in the 1960s the F-111 was deployed; the B-1 was developed during the 1970s, but the Carter Administration blocked its deployment; then, in the 1980s, the B-1 was resurrected and deployed as a part of a decision that also included undertaking the development of the Stealth B-2 bomber. More recently, initial design work has begun on the development of a hypersonic aerospace plane that can be used to deliver strategic nuclear weapons.

Along with this continuous incremental innovation of bombers with superior performance capabilities, the United States developed a variety of new vehicles. Beginning in the 1950s the ICBM was developed. Like strategic bombers, the ICBM has evolved through a series of generations beginning with the Atlas, followed by the Titan and Minuteman, and now being replaced by the MX. A similar pattern of innovation has occurred with intermediate-range ballistic missiles based in Europe. During the 1960s another innovation was added with the development of submarine-launched ballistic missiles. These systems have evolved from the Polaris through the Poseidon to the Trident. Finally, in the 1980s three forms of Cruise missiles—land-based, airplane-launched, and ship-launched—have been developed. As a result of these innovations, the United States now has seven different delivery vehicles.

This pattern of continuous innovation, with its associated innovation-to-obsolescence cycles, has been driven by Soviet efforts to counter each U.S. innovation with a comparable weapons system of their own. In general, the Soviets have played catch-up. Driven by a desire not to allow the United States to gain a significant qualitative advantage over them, and driven also by the hope that *they* might gain a qualitative advantage over the United States, the Soviets have built a huge and diverse organizational capability to carry out the synthetic innovation of military hardware. The arms race, then, has had the two superpowers continuously seeking to gain military advantage by out-innovating each other.

The initial focus on nuclear weapons and strategic delivery vehicles has, over the course of the Cold War, expanded to cover the whole spectrum of military technologies. Continuous innovation has created generation after generation of weapons systems, with each generation having superior performance characteristics. In addition, the organizational complexes that have produced these incremental innovations have also continually produced radical innovations. Today this process of continuous innovation includes weapons systems ranging from the rifles carried by infantrymen to spy satellites. To meet the need for continuous innovation and to protect against the consequences of obsolescence, the United States built an ever larger, more diverse, and more sophisticated organizational complex to carry out synthetic innovation. Initially labeled a military/industrial complex by President Eisenhower, this huge organizational system supports and utilizes nearly the whole range of scientific/technical activities.

COMMERCIAL SPINOFFS

As the military/industrial complex developed and grew, a large portion of the American manufacturing sector was integrated into it, as were a large set of government research facilities and the nation's research universities. Most of the industrial organizations involved produce not only military hardware and weapons systems but also products for the commercial marketplace. As the complex developed, a "spinoff" or spillover phenomenon began to occur with increasing regularity and speed. Spinoff refers to the transfer into the commercial marketplace of innovations or capabilities initially developed to meet the nation's military or space needs.[14] Obvious examples are the nuclear power plant and the 707 commercial jet. The first nuclear power plants were modifications of nuclear propulsion systems developed for Navy vessels, while the 707 jetliner was a modification of the Air Force's C-135. This spinoff phenomenon became pervasive. For example, plastics and composites that had been developed to meet military needs began to be substituted for metals and glass in the commercial sector. In no area was the spinoff of military innovations more evident than in electronics, because military aircraft and space vehicles had created a compelling need for highly reliable, miniaturized electronics systems. Thereafter the products and technologies developed to meet these needs spread into consumer electronics at an accelerating rate. For instance, electronic watches began to replace spring-driven watches; whole new product lines, such as electronic games, began to appear; communications satellites began to replace copper wire; and cardiac pacemakers became commonplace.

Although the pattern of continuous innovation had existed in some sectors of the electrical and chemical industries prior to World War II, the spinoffs from the arms race changed commercial products fundamentally. In those product areas where synthetic innovation capabilities already existed, the rate of innovation accelerated, and in those product areas not previously subject to continuous innovation, the impact was revolutionary. The range of products affected by spinoffs from defense innovations expanded steadily. With each new innovation-to-obsolescence cycle in military hardware and weapons systems, a new set of spinoffs occurred. Steadily and ineluctably throughout the post–World War II period, growing portions of the American commercial manufacturing sector began to take on the same characteristics as the defense sector; i.e., they became a part of the synthetic system. Competitive advantage flowed increasingly to those organizations leading in the innovation of commercial products. In contrast to the situation in the defense sector, however, in the commercial/industrial sector the growing importance of innovation to commercial success was not planned. The broader implications of what was occurring were little appreciated even in those commercial sectors that benefited most from spinoffs. Commercial innovation was seen as an essentially automatic and incidental development. U.S. leadership in the innovation of civilian technologies and commercial products was viewed as just another of the wondrous benefits of the free market. Certainly there was no perception of a need to focus the same kind of self-conscious, concerted national policy attention on commercial innovation that was required in the defense sector.[15]

At first the primary competitors of American producers of commercial high-tech products were other American producers. They were all linked to the same military/industrial complex. American companies were all sustained by the same process of arms race innovation, and they all shared the same business philosophy. It was a philosophy that held it is inefficient to replace a product that is selling well—or a process that works—with one that has superior performance capabilities simply on the basis of those superior performance capabilities. You did not build new manufacturing plants or open new production lines until the old ones were amortized or worn out. In sum, commercial innovation was not a dominant goal; rather, it was pursued when the opportunities and advantages became evident in the natural course of events.

Throughout much of the post–World War II period, then, the spinoffs of defense technology were making innovation increasingly important to the commercial-products sector, but commercial innovation was a secondary priority both nationally and for large numbers of companies. In the absence of foreign competitors such as the Soviets represented in the defense sector, there was little need to deliberately accelerate the time it took for an innovation-to-obsolescence cycle to occur. Americans simply got used to commercial innovation being automatic and incidental. These attitudes and expectations left the nation ill-prepared for the emergence of foreign competitors who had a "built-in" capability for continuously synthesizing commercial innovations.

THE EXPORT RACE

When other nations, led by Japan, began to develop capabilities for synthetic innovation focusing on commercial products, U.S. leadership eroded. Japan began to represent a challenge in the commercial sector analogous to that represented by the Soviets in the defense sector. Long before the United States recognized that there was an export race, it had been the first priority of the Japanese.[16] In its present incarnation, the importance of international competitiveness has made commercial synthetic innovation the highest-priority goal of Japan. The Japanese believe that high-tech exports are critical to their future well-being. Understanding the pattern of development Japan followed in building its capacity for synthetic innovation is important for two reasons. First, the pattern demonstrates the consistent focus of our most important competitor in the export race. Second, the Japanese model is being followed by a number of newly industrializing countries (NICs). If successful, these NICs will become increasingly serious competitors of the United States, not only in traditional manufacturing but also in the market for high-tech goods. The development of the Japanese model will be described in some detail in Part III.

Are Defense Spinoffs Enough?

In the 1980s the U.S. export position vis-à-vis the Japanese and other developed trading partners, as well as the growing group of NICs, remains heavily dependent on spinoffs from defense technologies. This point is emphasized by American leaders who support the Strategic Defense Initiative. General James Abrahamson, Jr., the first director of the SDI program, contends that it "will so stimulate the national economy that it will pay for itself."[17] Whether the United States can regain commercial high-tech leadership through spinoffs is the focus of vigorous debate. In the words of Paul Lewis of the *New York Times*, the Abrahamson view is supported by some economists who "argue that military research can develop new technologies that companies could never afford to finance. Others believe that on balance it weakens a nation's civilian industry by encouraging inefficient methods and absorbing scarce skills and resources that would be better used to produce goods for mass consumption."[18]

Those who doubt the future value of commercial spinoffs from defense innovations "argue that as modern weaponry becomes more sophisticated the technologies it requires are becoming increasingly disassociated from civilian needs."[19] This notion of divergence has become quite complex. The argument is that defense innovation has become so specialized that it cannot provide consistent and cost-effective support for the nation's commercial industries.[20] A common illustration is the military's emphasis on supersonic aircraft. Critics argue that contemporary military designs hold few lessons for the commercial aircraft industry. At a time when high performance in speed, maneuverability, and range continue to be primary goals of the military, the civilian market demands improved fuel efficiency and less noise.

The notion of divergence is also linked to a growing body of evidence that in rapidly moving civilian sectors the defense system is lagging. In some areas of advanced electronics, for example, defense systems are said to lag well behind civilian applications.[21] One explanation is that it takes the military so long to establish its very demanding specifications for electronics equipment that, by the time the specifications have been established and moved through the defense bureaucracy, consumer electronics products are using microcircuitry with performance and quality characteristics superior to those specified by the military. Some argue that the heavy integration of the U.S. electronics industry into the Department of Defense has gone some distance toward making that industry uncompetitive. With an assured government market, U.S. electronics producers tend to produce to military specifications rather than moving rapidly along developing state-of-the-art products in the commercial sector.

On two things there is a near consensus. The first is that government programs aimed at generating innovation in defense and medicine have been fundamentally important to the competitiveness of U.S. companies in the export marketplace. The second is that the pattern of indirect commercial or economic benefits—spinoffs—is no longer providing the United States with enough technologies that are competitive in the international marketplace.

CHAPTER 3

The Synthetic
Innovation Process

THE NEW REALITY rests on the capability to carry out continuous synthetic innovation. Synthesis involves combining information, knowledge, experience, and materials in ways they have never previously been combined to create products, processes, or projects with capabilities and characteristics not previously available. Synthesis, therefore, is as much an artistic enterprise as it is a scientific one. In their most advanced form, scientific enterprises are guided by bodies of theory that provide for rational understanding and management. Scientific theories identify key components or variables and explain cause-and-effect relationships among them. Thus those carrying out scientific enterprises know what variables to manipulate in ways that achieve their objective.

The process of synthetic innovation does not enjoy this kind of conceptual or theoretical understanding. Synthetic innovation requires experiential learning and trial and error as well as scientific knowledge and information. Learning how to synthesize has more in common with learning how to be an artist than it does with learning how to be a scientist. Furthermore, there are no textbooks on how to synthesize. Like art, synthesis is learned by doing, not by learning an ordered hierarchy of theory and information. Synthesis relies on knowing how to do more than we understand how to do, a characteristic underscored when the 1965 Nobel Prize was awarded to Professor Robert Woodward for his work in organic synthesis. Professor A. Fredga, member of the Nobel Committee for Chemistry, noted in his presentation speech that "organic synthesis is at the same time an exact science and a fine art."[1]

Artistic capability is a fuzzy concept, connoting a capability that cannot be objectively described or communicated in words or numbers. The ability to synthesize has the same quality. Like artists, creative organic chemists synthesizing a new compound, for example, know how to do more than they understand or can explain. The synthetic society is a vastly expanded version of the same phenomenon. It is dominated by organizational complexes that know how to do more than they understand and by organizations that have a great deal of experiential learning and a great capacity for carrying out trial-and-error tests.

Management consultant Robert Waterman, Jr., in his book *The Renewal Factor*, labels the process "informed opportunism." He emphasizes that innovation is characterized by uncertainty and randomness.

> The Ford Taurus, for instance, could never have been predicted up front. Initially, its proponents were sure about only two things: They didn't like the products Ford currently had on the market, and they wanted to build a better quality product. That was the extent of their "strategic plan." The car that eventually rolled off the assembly line was the product of thousands of little inventions and suggestions—from customers, engineers, workers, and designers.

> Random as they are, all the stories on innovation share an important characteristic. The person or team behind the innovation was both current with the relevant technology and had a feel for the market place.[2]

Components of Synthesis

Urea, the first man-made synthetic organic compound, was the creation of an individual. The most complex synthetic project ever proposed, the Strategic Defense Initiative, will be, if successful, the creation of an organizational complex. But whether synthesis is carried out by individuals or by organizational complexes, the generic process is the same. Recall that synthetic innovation involves the use of new ideas or methods. It involves the development and use of products or processes to solve some problem, meet some need, or to create and/or supply some market. Synthetic innovation is concerned with producing a product or process to carry out a task that has not been previously possible or, alternatively, with producing a product or process that allows a task to be carried out differently, at lower cost, or in a superior way. Keep in mind, too, that synthetic innovation occurs by combining information, knowledge, experience, and materials that have never been combined, or it occurs by combining these components in different ways. It is this new or different combination that requires the new ideas or methods.

The elements of synthetic innovation are nicely illustrated by an apocryphal story about a medieval blacksmith. One day the blacksmith was working at his forge when a knight rode up on a white charger, carrying a bent sword. The knight explained that he had gone out to save a damsel from a fire-breathing dragon. He had attacked the dragon, swinging the sword at its neck, only to have the sword bend. The knight asked the blacksmith to make a sword that would take off the dragon's head. The blacksmith replied that he did not know whether he could make such a sword, but, he said, he had found a clump of metal of a type he had never previously used. If the knight would leave his sword, the blacksmith would put the metal and the knight's sword in the forge, beat the metal into the sword, and the knight could come back the next day and try it. The following day the knight picked up the sword and once again attacked the dragon, with the same result: the sword bent. This process was repeated for four more days, with the blacksmith using different pieces of metal, until finally the knight rode up with the beautiful damsel and a bloody sword. The blacksmith, being an entrepreneurial fellow, immediately put up a sign that said, "Joe Doaks, Dragon-Slaying Sword Maker." He became world-renowned as the maker of dragon-slaying swords.

This apocryphal story illustrates the characteristics of synthetic innovation. First, the blacksmith was faced with producing a product capable of carrying out a task that had not been accomplished previously. The blacksmith approached this task by surveying the options available to him—specifically the availability of clumps of metal that he had never before used in sword making. He then proceeded through a series of trial-and-error tests. Specifically each day he beat a new kind of metal into the sword and the knight took it out and tried it. Finally, on the fifth day, a sword was produced that had superior performance characteristics, namely, the ability to take the dragon's head off. Having gone through the process, the blacksmith was from that time forth able to produce other dragon-slaying swords by following the established procedure. He did this by starting with what was previously a state-of-the-art sword and beating in different clumps of metal in a certain order. Thus the blacksmith created a new state-of-the-art capability.

It is important to emphasize that, although the blacksmith could now consistently produce reliable dragon-slaying swords, he did not understand what he had done. In the absence of that understanding, the blacksmith was unable to fully communicate in words and numbers how to make dragon-slaying swords. Nevertheless, the capability was not lost because the blacksmith could take on an apprentice who, by observation and participation, would be able to make dragon-slaying swords and therefore perpetuate this particular capability. The process of defining goals that are beyond the state of the art and pursuing them through trial-and-error tests, guided by both experiential and cognitive learning, characterizes the generic process of synthesis. This is the same process used by the organic chemist seeking to synthesize a new compound and the organizational complex seeking to synthesize a new project.

In each case, given an objective that is beyond the state of the art, the first step is to review what is known that may be relevant to carrying out the task. In the case of organic chemists seeking to synthesize a new molecule, they will review the existing literature that is possibly relevant to the reactions necessary to synthesize the desired molecule. Using that body of literature and their own experiences, they will then define a range of reactions that can be tried. The next step will be to run those reactions, starting with the ones they believe are most likely to produce the necessary compound with the desired yield. Success comes when the appropriate reaction is run and the desired compound in the desired yield is produced. The most successful organic chemists are those who, because of their accumulated experience, have a special feel or instinct that leads them to formulate and carry out the appropriate reaction. In highly complex synthetic projects such as the SDI, the same series of steps are involved, except that they must be carried out by groups of people located in networks of organizations. The synthetic process can be accelerated by defining a wide range of optional approaches and carrying out the trial-and-error experiments simultaneously rather than serially. That is exactly what the SDI is doing in its search for a workable system.

Complexity

Synthetic innovations range from the simple to the complex. The simplest synthetic innovations are those that can be conceptualized, formulated, developed, tested, and produced by an individual. Complexity increases as the quantity and diversity of knowledge, information, skills, experience, and materials necessary to produce a synthetic product or process increase. As synthetic products, processes, and projects become more complex, they pass beyond the capability of individuals to create, develop, and manage them. For example, a prototype of a synthetic organic compound may be produced by an individual chemist. On the other hand, a manned space vehicle like the space shuttle requires the contribution of thousands, if not tens of thousands, of people representing a diverse range of knowledge, information, and skills. As technology has become ever more complex and as we have come to use it to accomplish tasks that are further and further beyond the state of the art, technological innovation has come to require more and more complex organizational networks. Organizational complexes have become essential to carrying out complex synthetic innovation for a simple reason: the range of information, knowledge, experience, and skills needed to accomplish complex innovation is vastly greater than any single human mind can assimilate and manage.

Even a genius can understand or learn only a minuscule amount of the knowledge presently existing in the scientific/technical/production/deployment community. Information and knowledge have been exploding since the beginning of the modern scientific era. Derek de Solla Price has suggested that scientific information has continually doubled about every fifteen years over the three centuries of the modern scientific era.[3] Two hundred years ago it was possible for a well-educated person of intelligence to be roughly cognizant of the total corpus of scientific/technical information and knowledge. President Kennedy made this point memorably well at a dinner he held for Nobel laureates in the White House on 29 April 1962. In his after-dinner remarks the President said, "I think it's the most extraordinary collection of talent, of human knowledge, [ever to be] gathered together at the White House—with the possible exception of when Thomas Jefferson dined alone."

There are no Thomas Jeffersons today. The reason is simple. There is such a huge mass of scientific/technical/production/deployment knowledge, information, and skills that to assimilate and manage it has required dividing and subdividing knowledge and skills repeatedly. The major scientific and engineering disciplines, such as physics, chemistry, electrical engineering, and so forth, are divided into specialties and subspecialties and subdivisions of subspecialties. This process of subdivision continues apace. What is critical for scientific success or technical advance at the individual level is for the scientist, engineer, or technician to have state-of-the-art knowledge. Under the circumstances, it follows that, to have state-of-the-art knowledge, any individual must severely restrict the boundaries of the area she or he wishes to be expert in.

The genius of the synthetic society is its ability to create organizational arrangements that allow the expertise of the astrophysicist at MIT, for example, to be integrated with the skill of the tool and die maker in Los Angeles for purposes of making a space shuttle. Economist John Kenneth Galbraith has expressed this characteristic as follows: "The real accomplishment of modern science and technology consists in taking quite ordinary men, informing them narrowly and deeply, and then through appropriate organization arranging to have their knowledge combined with that of other specialized but equally ordinary men. This dispenses with the need for genius. The resulting performance, though less inspiring, is far more predictable. No individual genius arranged for flights to the moon. It was the work of organization."[4]

That the genius of the synthetic society lies in its organizational capabilities is difficult for Americans to fathom. As Galbraith observes, "The individual has far more standing in our culture than the group."[5] A common joke illustrates this. It is that a camel is a horse put together by a committee. In the synthetic society a camel is much more likely to be a horse put together by an unspecialized, unqualified individual. This is the reality—only the committee can bring all the needed capabilities to bear. Our myth system, however, continues to perpetuate the belief that committees are barriers to creativity and that individuals are the source of technological creativity.

The size and character of the organizational complexes needed to carry out continuous synthetic innovation of complex products, processes, and projects is determined by the need for two types of integration. One involves the integration of knowledge and skills at different points in the innovation process, distributed along a continuum that runs from fundamental knowledge to application (use). The second type of integration runs along a continuum that divides knowledge, skill, and information by substantive area of specialization.

It is useful to conceptualize the body of knowledge, skill, and information needed to carry out a complex synthetic innovation as a matrix (see figure 3–1). Along the ordinate (horizontal) axis the first level is basic or fundamental scientific knowledge and information. As we move to the right, the next broad category is applied science, then engineering, followed by a set of technical, production, and marketing or deployment skills. For example, one row of the matrix might begin with theoretical physics, proceeding to solid-state physics, then electrical engineering, followed by electronics technical work, then by pro-

FIGURE 3–1
Expertise Matrix

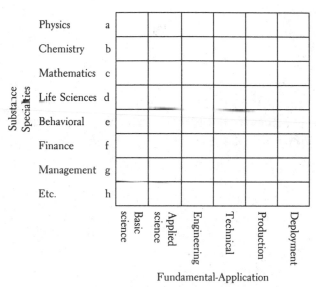

Fundamental-Application

(Complexity increases as the number of areas
of needed expertise increases.)

duction and marketing or deployment skills. The vertical side of the
matrix is divided according to the fundamentally distinct science and
business disciplines. The complexity of a synthetic innovation is determined by the total number of cells of expertise that are required to carry
it out.

Examples of the Innovation Process

Different levels of complexity can be illustrated by looking at three
different innovations. The first example is the "E-coat" painting process developed in 1958 by the Ford Motor Company development
manufacturing people. Their goal was to improve the quality of paint
jobs, particularly the rustproofing of the underbellies of cars and trucks.
Previous techniques for having paint reach all of a car's nooks and

crannies had yielded uneven coverage, leaving cars vulnerable to rust. In his widely read book *The Reckoning,* David Halberstam describes the innovation:

> In effect it was like painting a car the way metalworkers plated metal with silver or chrome. In the process, the car body was completely submerged in a tank of paint and given an electrical charge; the paint received the opposite charge. Thus the paint, electrically attracted to the car, was pulled into the tiniest, hardest to reach crannies of the body. From the first the process was a stunning success.[6]

The range of knowledge and skills involved in carrying out this innovation was not very large, since the principles associated with positive and negative charges were widely understood. Thus the innovation could be carried out by people with applied expertise, located in the engineering/technical/production range of the fundamental-application spectrum (horizontal axis in figure 3–1), in response to market demand. It was a relatively simple innovation, but one that represented a significant improvement in quality and performance.

The second example is the microchip innovation. Here the goal was to develop a much smaller, more reliable, and faster way to manage electronic data in response to defense and space needs—in other words, to develop a higher-performance substitute for the transistor. Developing the microchip required tapping knowledge in the solid-state physics area about the behavior of atoms in solids. It also required the fabrication or synthesis of the most appropriate materials. In this case, then, the needed integration or combination involved a range of skills broadly distributed along the fundamental-application spectrum as well as information, knowledge, and skills from several different substantive areas of expertise. Once a prototype with the desired performance characteristics had been developed and tested, other skills were needed to develop the capability to manufacture chips on a mass basis at an acceptable cost.

The third and most complex example is the SDI, which requires knowledge, skills, and information ranging from the most fundamental to the most applied in nearly the full range of substantive specialty areas.

Successful synthetic innovation involves more than simply integrating discrete bodies of knowledge, information, and skills. It requires combining them in ways that deliver synergism. That is, innovations are more than the sum of the bits of specialized expertise of which they are composed. Thus the synthetic-innovation process delivers something new; it is a creative process. In addition, it is important to emphasize that synthetic creativity can range from big new ideas to elegant new combinations of existing things. Management consultant and author Robert Waterman illustrates the latter with the example of the Ford Taurus. The Ford team, as part of its design effort, selected a dozen of the best competitor cars (midsize, four-door) and subjected them to a systematic disassembly. The team learned a lot. Specifically, they identified 400 features that went into the Taurus "wish list." Ford reckons it met or bettered the competition on 360 of the 400 items.

> In isolation . . . it sounds like copy cat stuff. But Ford people will tell you that this is the first time in a long while they haven't felt like imitators following a me too strategy.[7]

The innovation came from the synthesis of already existing features and technology.

Driven by the need to accomplish tasks, innovation in the synthetic society is a planned and not a random activity. It is a creative process that requires integration carried out for the deliberate purpose of producing in a synergistic way new or superior products, processes, or services. Thus where the purpose of innovation is superior performance—as is the case in defense and medicine—with cost being a secondary concern, cost will be sacrificed to performance. Similarly where the goal is product performance, production processes will receive little emphasis. In another variation, where the goals of innovation include lower costs, higher quality, and production simplicity as well as superior performance, the synthetic process can yield miracles. Roland Schmitt, president of Rensselaer Polytechnic Institute, suggests the important role of the goals sought by innovation. Pursuit of these goals begins with the design phase.

In the U.S. the design phase of the cycle of innovation concentrates on the features and performance of the product much more heavily than on the processes by which it will be manufactured. We design the product first then tackle the job of how it is made. The U.S. designer, starting with a certain trade off between cost and performance, proceeds through subsequent design iterations to add features and performance and ends up with higher cost. The Japanese designer designs to cost; however many design iterations there may be, the initial cost barrier remains inviolate.[8]

Keep in mind that whether the goal is efficiency, quality, or performance, the key to success in the synthetic society is to get there ahead of or at least at the same time as your competitors. In combination, the importance of time and the disorderly experiential character of synthetic innovation requires that all of the organizations that make up a given complex share common goals and have great flexibility. As the Congressional Office of Technology Assessment notes,

Despite the common perception, processes of technology development seldom follow a linear progression from research, to development, to commercialization, and production. Sometimes new knowledge in the form of research results or specialized expertise that may not be widely diffused within the technical community is needed at the beginning of a development project (developing a new product or process often takes five years or more). In other cases such knowledge will prove critical near the end of a project. Typically, research enters the process at multiple points. Moreover, the great majority of American firms, large as well as small, lack the resources to be self-sufficient in technology and science. The complexity and pace of modern technology have simply outstripped their ability to keep up. Institutional mechanisms for technology development and diffusion—transferring know-how in research results to those in industry who need them—must not only be flexible, they must be capable of responding to widely varying needs at different stages in development and commercialization."[9]

The importance of having the diverse science/technology/production/marketing organizations continuously and intimately linked with each other is further underlined by the fact that successful technology development frequently rests on what some have called "fingertip knowledge"; that is, it rests on transferring knowledge and skills that cannot be communicated in words or numbers. It is a given within the R&D community that the most effective route to technological innovation is through the interaction of people. This occurs only when the boundaries among different organizations are highly permeable. In the United States that permeability has characterized organizations within the military-industrial complex but is much less the case in the commercial sector.

In the United States, scientific-technical production and marketing or deployment expertise are distributed among a wide range of diverse organizations. It is common to divide them into six categories: government agencies, government laboratories, nonprofit organizations, federally funded research and development centers (FFRDCs), universities, and industrial organizations. In general those who discover or create the largest portion of fundamental knowledge—the basic researchers—are located in the nation's graduate research universities. Alternatively, the vast majority of the capability for applications (production or deployment) is located in industrial organizations. Government agencies normally play the role of funding and providing direction for complex innovation projects in defense, medicine, and agriculture. In addition, these agencies play the critically important role of establishing and facilitating linkages among all of the other organizations. Government laboratories, nonprofits, and federally funded research and development centers play a wide range of different roles and contain within them a diverse mix of expertise.

Within all six of the different types of organizations, expertise tends to be organized around substantive activities or substantive areas of expertise. For example, in universities expertise is organized by scientific and engineering disciplines, while in nonprofits, FFRDCs, and government laboratories, expertise is likely to be organized around applications areas or fields such as materials, acoustics, electronics, and optics. In industrial organizations, expertise is more likely to be organized by product or process. The government agencies that support, coordinate, and facilitate the linkage among these different organizations are normally made up of specialized units that focus on everything from basic research to generic problem areas to specific products or processes.

Complexes and Networks

It is useful to divide into two types the organizational arrangements that are central to synthetic innovation in the United States: *complexes* and *networks*.

A complex is composed of all the organizations involved in a general area of activity or product sector. For example, the military-industrial complex includes all the organizations that participate in defense innovations: (1) defense agencies (e.g., DOD, DOE); (2) federal laboratories (e.g., Wright-Patterson, Redstone Arsenal); (3) FFRDCs (e.g., Lawrence Livermore, Sandia); (4) nonprofit (e.g., RAND, MITRE); (5) universities (e.g., MIT, Johns Hopkins); and (6) industry (e.g., Lockheed, Texas Instruments). A complex is the "holding company" for the expertise needed to carry out the range of specific innovations that produce a given product. In addition, individual organizations may be participants in more than one complex. To be part of a complex minimally requires that the organization be involved in the innovation process of a sector; for example, General Electric is involved in both the defense and medical sectors.

A network is organized around a discrete activity, product, or process. Networks are subsets of an organizational complex, composed of that specialized expertise and capability needed to carry on a specific activity or to deliver a specific product or process. For example, all of those people in and components of organizations directly involved in and/or funded by the SDI Office (SDIO) are part of the SDI organizational network.

The Synthetic Innovation Process

Organizations may enter or depart both complexes and networks through either external or internal initiative. For example, DOD may bring a university into the military/industrial complex by funding professors to do research in general areas thought to be relevant to future defense innovations. Alternatively, an organization may be brought into a network by an agency issuing a contract for work that is needed to carry out a specific innovation, or a company may decide to enter or abandon a particular business area. The key point is that participation in either complexes or networks requires relevant expertise or capabilities. In contemporary society the frontiers of science, technology, and production are assumed to advance continuously, and any organization that does not stay abreast of the state of the art is obsolete. Thus possessing relevant expertise and capabilities requires continuously pushing the scientific/technical frontier. For example, in the defense complex, federal R&D contracts pay organizations to stay abreast of and/or push past the expertise frontier. Follow-on production contracts and future R&D contracts provide additional stimulus for organizations to push the state of the art. On the other hand, no similar broad-based pattern of funding and stimulus exists in the commercial high-tech export sector, for it is not seen as appropriate for government to support commercial product development directly. Government's role in the commercial sector is, with few exceptions, limited to funding and stimulating fundamental research. Indeed, initiatives and support for specific commercial product and process innovations come primarily from industry and, in limited cases, from various R&D organizations. These initiatives are normally product-, process-, or activity-specific. In general, the corporation or the R&D organization pushing the commercial innovation acts to build and, where successful, to sustain the network of organizations necessary to carry out the innovation(s).

When one compares the defense complex with the commercial high-tech complex, two things stand out. First, no agency, organization, or set of organizations has responsibility for insuring the health and vitality of the commercial high-tech complex. Specifically no organization plays the role the Department of Defense plays for the military/industrial complex. Second, for political/ideological reasons government laboratories, FFRDCs, and government "captive" nonprofits (i.e., those solely funded from public monies) play a very limited role in the commercial high-tech complex except where the commercial innovations are direct spinoffs of defense technology.

Organizational Learning

Given the objective of producing products or processes that are beyond the state of the art and that require synthesizing a wide range of expertise and skills, and given a lack of formal understanding of how this is done, how do organizational complexes accomplish innovation? They rely heavily on the phenomenon of "organizational learning." The concept of organizational learning refers to the capacity of organizational complexes to develop experiential knowledge, instincts, and "feel" or intuition which are greater than the combined knowledge, skills, and instincts of the individuals involved. They have the capacity to learn whatever is critical to their ability to innovate continuously. Thus the organizational complexes that synthesize learn in ways similar to the kind of learning demonstrated by the medieval blacksmith described earlier. That is, they know how to carry out synthetic innovation even though they do not understand it. Robert Waterman calls this the "interaction factor." He also notes that H. Ross Perot refers to this as knowing your business. "It means being able to bring to bear on a situation everything you've seen, felt, tasted, and experienced in an industry."[10]

Two points are central to appreciating that innovation is a product of groups that can do more than they understand. First, individual experts are not simply interchangeable parts in an innovating machine: innovation involves team building and continuous team learning. With each cycle of innovation, that team capability is enhanced. In his *Tales of a New America* Harvard's Robert Reich insightfully characterizes the process as follows:

> Individual skills are integrated into a group whose collective capacity to innovate becomes something more than the simple

sum of its parts. Over time, as group members work through various problems and approaches together, they learn about each others' abilities. They learn how they can help one another perform better, who can contribute what to a particular project, how they can best gain experience together. Innovation is inherently collective and incremental. Each participant appreciates what the others are trying to do; he is constantly on the lookout for small adjustments that will speed and smooth the evolution of the whole. The net effect of many such small-scale adaptations, occurring throughout the organizations, is to propel the enterprise forward. Because such cumulative experience and understanding is so critical, this network of people must be maintained over time. Their collective capacity, by assumption, cannot be translated into standard operating procedures and transferred to other workers. The capacity to add value resides in the whole, and this capacity is what the enterprise seeks to preserve and develop. This is in sharp contrast to the pattern in standardized production, in which drone workers are seen as interchangeable, and extruded when the industry or product approaches the end of its life cycle.[11]

Second, there is a critical need to compete in the innovation process continuously. Much of the success in innovation in the defense organizational complex flows from the perception that the United States must never lose its lead in innovation. Failure to enter the innovation competition early for specific areas of technology and, more important, failure to remain a continuous innovator is costly. The cost flows from a failure to experience the continuous organizational learning curve that parallels each incremental innovation. Where experiential organizational learning is critical to innovation, failure to participate in any cycle increases the costs and difficulty of seeking to reenter the competition later.

Akio Morita, president of the Sony Corporation, provides insight on the costs of dropping out of the innovation process:

We later marketed a special calculator model we called SOBAX, which stood for "solid state abacus." But I soon realized that several dozen Japanese companies had jumped into the business of making calculators, and I knew the shakeout would come sooner or later through a very brutal price war. That is the way it is on the Japanese market, and it was just the kind of thing we have always wanted to avoid. When it became obvious that others would be discounting dangerously to get a share of the market, we gave up the calculator business.

My prediction was right. Many calculator makers went bankrupt and others just got out of the market, taking a big loss. Today there are only three major makers of calculators, and in a way I have been vindicated. There was still much to be done in audio, television, and video to keep us challenged, and we were always looking for new applications.

But I must say here that, on reflection, I was probably too hasty in making the decision to get out of calculators. I confess that today I think it showed a lack of technical foresight on my part, just the thing I think we have been good at. Had we stayed with calculators, we might have developed early expertise in digital technology, for use later in personal computers and audio and video applications, and we could have had the jump on our competition. As things developed, we had to acquire this technology later, even though we once had the basis for it right in-house. So from a business viewpoint we were right in the short term, but in the long term we made a mistake.[12]

In the United States, government has assured continuous organizational learning in the networks and complexes that carry out innovation in defense, medicine, and agriculture. No similar continuity has existed in the commercial sector, however. We have not done what Robert Reich says occurs in Japan:

The emerging Japanese-American [Reich's label for the large Japanese companies] corporation allocates to the Japanese the most important asset for the future—experience in making complex products cheaply and well. They learn how to organize themselves for production—integrating design, fabrication, and manufacturing; using computers to enhance their skills; developing new flexibility; creating new blends of advanced goods and services.

They learn how to make the kinds of small, incremental improvements in production processes and products that can make all the difference in price, quality, and marketability. In short, they develop the collective capacity to transform raw ideas quickly into world-class products."[13]

COLLECTIVE DECISION MAKING

Complex innovations must be formulated and tested by groups and teams whose decisions are reflected in consensus actions and recommendations. Consensus recommendations by one group or committee move to other, differently configured, groups or committees, and then to others. This is part of the organizational learning process—each group consists of people with different kinds of in-depth expertise. Individuals do not make a comprehensive decision, not even those individuals who hold positions of organizational authority. Alternatively, every expert has the right to say what is beyond the state of the art in her or his narrow area of expertise. Options for accomplishing innovation are formulated and defined through a process of continuous information exchange and melding, often in committees. These options fall within a range of capability defined by the experts who make up the groups.

It is important to emphasize that in the organizational networks that innovate complex technologies, the groups and teams critical to the process are not just made up of people from single organizations. For example, the teams or task forces that manage the innovation of space technology are not made up only of NASA or Lockheed employees. Rather, the groups are comprised of people from whatever organizations have the expertise considered necessary to formulate and test the specific options. If the innovation is complex, it requires integrating expertise that ranges from basic research to deployment or marketing.

ORGANIZING FOR LEARNING

In the United States, as we have seen, continuous innovation in specific product or process areas rests on networks that make it possible to tap and organize specialized talents wherever they may be located. Networks and complexes, not individual organizations, are the repositories of the organizational learning that is necessary to carry out synthetic innovation. It is the accumulated learning of the defense complex, for example, that provides the basis for the decision to undertake the SDI. The strong record of the United States in accomplishing complex innovations is what makes the SDI a credible threat to our Soviet adversaries. One of the striking conundrums of our time is that some of the most vocal and probably most effective critics of the SDI are scientists and engineers who understand its great complexity and its need for synergism and, based on that recognition, believe it will not work. This reaction to the SDI is hardly a first. One is reminded of the conclusion of Vannevar Bush, who headed the Office of Scientific Research and Development during World War II and was one of the great seers of the coming technological society. Despite his visionary qualities, he concluded "that there would never be such a thing as an intercontinental ballistic missile."[14] Bush was proved wrong, soon after he made the statement, by the very organizational complex he helped build.

The capacity for organizational learning makes U.S. defense efforts such as the SDI particularly threatening to the Soviets. Since the critical organizational genius is, in part, noncognitive and only exists where organizations have accumulated experiential learning, it poses a unique problem for the Soviets. Not even the most effective spy network can tap into and communicate this experiential learning. Thus the Soviets fear a quantum jump in U.S. capability, which they can draw on only after it has already been built in the United States.

In the commercial sphere, the same kind of organizational learning and experience is what has given the Japanese a significant advantage in innovating superior manufacturing capabilities. David Halberstam describes this process as follows:

> The Japanese had moved ahead of America when they were at a distinct disadvantage in technology. They had done it by slowly and systematically improving the process of their manufacturing in a thousand tiny increments. They had done it by being there on the factory floor as the Americans were not.[15]

The Synthetic Innovation Process

Halberstam characterizes the view of Professor Harley Shaiken, a labor economist who has studied this area, as follows:

> The only way to learn, he believed, was to do it. To struggle through and to make endless mistakes, define options and subject them to trial-and-error tests, but as you did so constantly improve your process and your workers. That was what the Japanese secret had been.[16]

Whether one is concerned with achieving small, incremental innovations in process, as characterized by the Japanese success in manufacturing, or radical new project innovations, such as the SDI, the generic characteristics are the same. These innovations require organizational systems that both allow and stimulate the conceptualization of new options and that provide for trial-and-error efforts aimed at synthesis of products, processes, and projects with new or superior performance capabilities. An important element of the continuous innovation process is the ability of organizational complexes to recognize and exploit serendipitous discoveries. It is the development of these capabilities that has made innovation the critical ingredient of competitive leadership in the contemporary period.

The Challenge of Synthetic Innovation

The preceding chapters have emphasized that U.S. leadership in defense, medicine, and agriculture derives from these sectors having built organizational complexes specifically for the purpose of developing products and projects that are beyond the state of the art. The fundamental distinction between the United States and Japan has been the differing foci of the innovations for which organizations with these synthetic capabilities have been built. The Japanese have focused very heavily on the innovation of commercial products, and they have focused major attention on process innovations, that is, manufacturing

processes that deliver products of higher quality, with higher performance characteristics, and at lower cost.

As the Japanese focus on radical innovations, their critical challenge
is organizational. Radical innovations tend to rely heavily on the ability
to draw on and integrate information and skills ranging from basic
research to the most applied activities. To date, Japanese synthetic
innovation has generally been possible using personnel located predominantly within single companies. As the Japanese move to innovations
that are further and further beyond the state of the art—for example,
the "fifth generation" computer—the quantity of specialized skills
required increases markedly. Inevitably this synthetic innovation will
require tapping fundamental knowledge in mathematics, physics, and
materials, and so forth. Much of that knowledge is located in universities and organizations outside the corporations and outside Japan. The
question facing the Japanese as they move into a major focus on radical
innovations is whether they can develop the flexible organizational
arrangements needed to achieve such innovations.

One point is absolute. In the absence of bodies of theory or conceptual systems that provide understanding of how synthetic innovation occurs, such innovation will rest on experiential learning, trial-
and-error tests, and serendipity. Therefore, contemporary, complex
synthetic innovations have a common characteristic: no matter what
society they occur in—whether Japan, the Soviet Union, the United
States, or anyplace else—there is an irreducible requirement for
organizational arrangements that can at once draw on and integrate
expertise and skills ranging from basic research to marketing and also
achieve synergism.

Nothing is more critical to the future competitiveness of the United
States in the export market than a broad appreciation of what is
involved in complex synthetic innovation. That capability was initially
developed in the United States and today exists in its most advanced
form here in defense, medicine, and agriculture. The failure to transfer
the lessons learned from these three sectors to the commercial export
sector is a source of our export weakness. That failure is in large part
the result of self-deception, deriving in turn from a myth system, an
ideology, developed during the nation's industrial period. It is to that
myth system that we turn next.

CHAPTER 4

The Secular Trinity

T HE INDIVIDUAL, the free market, and efficiency are such fundamental American socioeconomic reference points that it is appropriate to label them the "secular trinity." The secular trinity is at the core of a contemporary American idea system about business that the capacity for continuous synthetic innovation has made an increasingly costly myth system. It posits that opportunity and achievement throughout the society are maximized by allowing individuals to compete with each other in a free market and that the winners in this competition will be those who are most efficient. The secular trinity provides a framework for organizing and managing society that Americans believe maximizes their benefits and minimizes their costs.

Although modern philosophy has devoted major attention to the importance of distinguishing between facts and values, ideologies are successful precisely because that distinction is not made. The power of the secular trinity derives from the fact that it provides an understanding of how reality works and why it does or does not work well at every level from micro to macro, from the individual level to the international level. In addition, it offers an operational value structure to guide both individuals and society in what should be done when we are dissatisfied with how reality is working.

Society succeeds, the trinity says, when creative, competent, hardworking, risk-taking individuals can operate in an environment that does not constrain their pursuit of efficiency. Alternatively, failure is the result either of flawed individuals or of an environment that constrains individuals by not allowing the most efficient to win. Faced with an unsatisfactory reality, the answer is to increase the performance of individuals either by improving their competence and/or motivation or by changing the environment in which they operate. The critical environmental changes are those providing greater assurance that individuals who are efficient will receive rewards, while individuals who are inefficient will not.

In sum, in the United States the secular trinity is the lens through which we see and understand reality and is the standard that guides our decision making and management. It is also an illusion. Reality is changing, but the secular trinity is not. Increasingly it is providing a distorted view of reality and is prescribing actions that are inappropriate to the nation's problems and needs.

The Mystique of the Individual

In the American social paradigm, the individual has a role analogous to that assigned to the atom by physicists in the early part of this century. That is, the individual is the indivisible building block of society. Individuals are where you start if you want to understand how society operates. Unlike atoms, however, individuals are not uniform. Quite to the contrary, some stand out; it is those outstanding people whom the secular trinity celebrates.

The American success story is commonly construed as the story of exceptional individuals. It began with those who left Europe pursuing religious freedom, and it has been carried on by a varied collection of heroes including George Washington, Abraham Lincoln, Thomas Edison, Henry Ford, Alfred Sloan, Edward Teller, the astronauts, and H. Ross Perot. American success is due to "rugged individuals" who crossed the frontier or broke the mold.

Belief in the secular trinity tells us to look for the exceptional individual when we want to understand why things work well and to look for exceptional individuals who can be put in charge when things are not working well. As emphasized by Lester Thurow, dean of MIT's Sloan School of Management, American mythology has long celebrated the Lone Ranger.

> The interesting thing about America's love affair with the Lone Ranger myth is not that the Lone Ranger did not in fact exist. It is that he could not have existed. The American West was not settled by Lone Rangers, precisely the opposite. It was settled by wagon trains and community barn raisings—social organization. Individuals alone on the high plains of Montana in 1840 or 1870 weren't successful—they were dead.[1]

In contemporary society, where a very large number of our products and services are delivered by large organizations, it might seem more difficult to attribute success to exceptional individuals, to "Lone Rangers." In practice, however, that has not been at all difficult; it has only required some modest redefinition of old categories. According to the secular trinity, organizational success in the contemporary period is a result of two kinds of distinctive individuals: entrepreneurs and managers.

ENTREPRENEURS

Entrepreneurs are pictured as eccentrics characterized by quick insights, quick action, a willingness to take risks, and tenacity. Entrepreneurs are the people who push the frontiers of society and the economy by developing new products and identifying new markets. They are the creators of new organizations that grow to be large organizations. They are the creators of new jobs.

No theme is more pervasive today than the notion that our trade problems require more entrepreneurs, especially technological entrepreneurs. Thus chairs and programs of entrepreneurialism are springing up in American business schools, and government is seeking to create a climate favorable to entrepreneurs. Large corporations are searching for ways to free and stimulate their exceptional employees. A new word has been coined for this type of individual; it is "intrapreneur." Even the government offers bonuses to its star civil servants, since the way to achieve efficient government is thought to be by providing individual civil servants with incentives similar to those offered by the free market.

Given the examples of Stephen Jobs and Steven Wozniak, who started Apple Computer, and H. Ross Perot, who started Electronic Data Systems and went on to become a major thorn in the side of General Motors, the entrepreneur is now commonly perceived as the source of high-technology products and services and the answer to the high-tech trade problem. In actuality, this is an illusion. It reflects a fundamental failure to understand what is involved in synthetic innovation. For example, Edwin Land, the inventor/entrepreneur who started and ran the Polaroid Company, is one of a declining number of technological entrepreneurs—i.e., individuals who achieved a major technical innovation and then successfully managed a company to develop and market products derived from it. Today this breed is almost extinct and the reason is simple. It is not because of some flaw in the training, imagination, or drive of individuals. It is because innovating complex technologies requires capabilities, competence, experience, and skills that are beyond the capacity of any single individual. Henry Ford could understand every step in the production of his Model-T. The complexity of most contemporary technologies precludes such individual innovation.

We have passed beyond the era of the inventor/entrepreneur. Whether it be videocassette recorders, digital watches, hypersonic airplanes, or "smart" weapons, it is no longer possible for individuals to invent or synthesize complex technological devices. In those rare instances where an individual is able to conceptualize and innovate a new product, it is likely to be a one-time, big-idea event. The invention, let alone innovation, of high-tech products* requires the ability to tap, integrate, and gain synergism from a range of knowledge, information, and skills that go far beyond the capabilities of a genius. Therefore, looking to individuals for invention and successful innovation of high-technology products and processes is a formula for disappointment.

Because of complexity the role of entrepreneurs in the high-tech product area will likely be the identification of market niches where

---

*Recall that innovation includes not only the discovery and invention phase but also the development, production, and marketing of new products and processes.

existing technologies can be used or sold. The genius of Electronic Data Systems' founder H. Ross Perot and Apple Computer's Steve Jobs was their ability to see a market for technologies that involved modifications of existing technologies. The future entrepreneurs of high technology are likely to be people who acquired experience within the defense, medical, or agricultural complexes and who know how to tap the accumulated organizational learning of those complexes to satisfy some newly perceived market niche.

The critical role of group/organizational learning to contemporary high-tech entrepreneurs is suggested by the tendency of electronics entrepreneurs to be located in Silicon Valley or along Boston's Route 128. These locations provide concentrations of expertise and organizational learning that entrepreneurs who see a new market can rapidly use to build their own network for the purpose of innovating products or processes. Unfortunately, many if not most such entrepreneurial organizations are doomed to fail because of their inability to carry out continuously the rapid incremental innovations demanded by innovation-to-obsolescence cycle competition. Innovation-based competitive success requires moving new ideas and methods to the market rapidly. Continuous innovation requires the resources to build and continuously change production and marketing capabilities as necessary to remain competitive in the international market.

The resources required to maintain and evolve innovative capabilities are considerable. Innovation-to-obsolescence cycles frequently drive technologies through boom and bust periods. When they are in a bust period, American corporations normally cut back on expenditure and, more important, on personnel. This is especially true of small entrepreneurial companies with limited product lines. Cutbacks mean that companies break up teams in which reside organizational learning that is likely to be crucial to future success. Thus size and continuity are critical. For example, it may actually be that the funding of entrepreneurs through the venture capital market has damaged the semiconductor industry by financing its disintegration into suboptimal small firms.

The belief that entrepreneurs will generate competitive new technologies that will save the day for America is a reflection of our failure to see the new reality. Very few contemporary technologies have an individual inventor. Competitiveness requires building teams, organizations, and networks that exist for the purpose of continuous innovation. It requires organizational systems that make innovation routine—not individual entrepreneurs.

MANAGERS

It is part of the conventional wisdom that the genius of the entrepreneur is his or her ability to start successful companies. But, with rare exceptions, those characteristics that allow for launching a company are the very ones that preclude successful long-term management. The normal formulation is that the instincts for quick insight, quick movement, and risk taking are at odds with what is necessary to provide predictable, sustained improvements in efficiency and profitability. For stable, rational, and, most important, efficient long-term management, our society looks to the professional manager.

Underlying the notion of the professional manager is the belief that there is a set of management principles that can be learned and that are applicable to all organizations. According to contemporary mythology, managers are the people who run large organizations and make key decisions. Successful managers are those whose companies grow, turn in steadily increasing profits, and have highly valued stock. Over the last two decades, large numbers of managers have come from that ubiquitous post–World War II development, the MBA program.

Managers—certainly those produced by MBA programs—are people who have purportedly been trained and equipped with concepts and methodologies that allow them to maximize efficiency, which is obtained by squeezing out waste and inefficiency. Management rules are relatively simple and straightforward: managers set progressively higher performance goals for their subordinates and, in relative terms, they reduce the resources available to achieve those goals. Managers succeed based upon the performance of lower-level managers.

The route to efficiency requires both the right kind of organizational arrangement and the right kind of people in charge. Thus modern managers—whatever organizations they may be in—begin by dividing the organizations into distinctive, clearly defined profit (cost) centers. That done, responsibility for the profit centers can be clearly identified with the managers. These arrangements allow successful management to be rewarded and failure to be dealt with rapidly. The performance of profit or cost centers, and thus of their managers, can be assessed over relatively short periods of time—e.g., quarterly—based on the success or failure of that now nearly hackneyed standard, the "bottom line." Successful managers are in charge of profit centers that consistently turn out adequate profits, the ideal being consistent incremental increases in profits quarter after quarter and year after year.

Thus central to good management in the United States is the clear identification of individual responsibility. This is an extension into the organization of the notion of individualism. You keep those who perform and get rid of those who do not, and performance is measured in the short term. Akio Morita of Sony articulates a diametrically opposed philosophy.

But the point is that these individual mistakes or miscalculations are human and normal, and viewed in the long run they have not damaged the company. I do not mind taking responsibility for every managerial decision I have made. But if a person who makes a mistake is branded and kicked off the seniority promotion escalator, he could lose his motivation for the rest of his business life and deprive the company of whatever good things he may have to offer later. If, on the other hand, the causes of the mistake are clarified and made public, the person who made the mistake will not forget it and others will not make the same mistake. I tell our people, "Go ahead and do what you think is right. If you make a mistake, you will learn from it. Just don't make the same mistake twice."

Besides, even if you find the person responsible for the mistake, I told my American friend, it is likely that he has been with the company for a while, and even if a replacement is put in his job, it won't necessarily make up for the loss of his knowledge. . . .[2]

If efficiency is the result of good management, it is appropriate for good managers to be promoted rapidly. If promotion takes the manager from hotels to bread to aerospace, that's no problem. It is not knowledge of the particular activity being managed that is important; it is the application of the universal rules of management. Successful management does not require in-depth, substantive knowledge. The idea is widely held that moving managers every two to four years is not only acceptable but beneficial. Managers left in the same location for long periods of time tend to grow too close to their fellow workers. They may become addicted to the promotion of specific long-term projects and lose sight of the central purpose of management, which is efficiency, measured over short periods of time using the bottom line.

The notion that trained and experienced individuals can make the key decisions necessary for an organization to be successful has become so powerful that it now pervades nonprofit as well as profit-making organizations. Even government has increasingly sought to apply management principles to its own operations. Though it does not apply a "bottom line" profit standard, government has sought to construct equivalents. Indeed, one thing the rhetoric of government management shares with business is that the goal is efficiency. Anyone who has spent time in government knows that there is a perpetual concern with strategic plans and performance objectives and evaluations. The same phenomenon now pervades universities, church organizations, and other nonprofits.

The belief in management as a profession rests on the notion that it is cognitively possible to understand how large organizations operate and how they operate best. With such cognitive understanding, organizational activities can be quickly communicated to and grasped by the continuing cycle of new managers brought in from the outside. The belief in a cognitive understanding of management runs counter to the synthetic reality, however. This is because organizational learning involves developing knowledge and capability that is greater than the sum of knowledge possessed by the indivduals in the organization. It is knowledge that is partly noncognitive. In the synthetic reality, decision making must occur by consensus because only decisions informed by a broad range of narrow expertise can lead to synthesis. There is no greater danger to an organization's success in the synthetic society than for managers to substitue their decisions for a substantively informed *consensus* decision making process.

The need to integrate diverse expertise plus the need for experiential organizational learning, two elements integral to the process of innovating complex technologies, preclude individual decision making. So, in the synthetic society, management—that is, detailed decision making by individual managers—is an illusion.

In the synthetic society, as we have seen, organizational success rests on continuous incremental innovation. In the United States, such innovation requires linking organizations and establishing organizational networks with effective communication systems. Harvard's Robert Reich insightfully describes, in the context of individual organizations, what is also needed in the organizational networks. "Because the information and expertise are dispersed throughout the organization, [in the best organizations] top management does not solve problems nor set specific direction. It creates an environment in which people can identify and solve problems for themselves."[3] More broadly, in successful synthetic organizations top management creates an environment in which new ideas can be tried and experimented with on a continuing basis. Through that process the organizational system gains the experience that is critical to synthetic innovation.

The distinctive individuals in the synthetic society are neither inventors/entrepreneurs nor managers per se; they are individuals who can in some way motivate and lead large numbers of people in complex organizational systems to accomplish what has never been accomplished before. Such individuals are rare. Although great leaders are valuable they are not necessary for success in the synthetic society. The continuing success of the IBM Corporation is a striking illustration. At any point in time, few people know who the president of IBM is. In truth, IBM prides itself on the fact that it has available adequate replacements for any job in the company, including the chief executive officer.

The key to success in the synthetic society is the team and its capacity to build so much innovation capability that the organization evolves and innovates continuously. Robert Waterman makes the following observation:

American emphasis on individualism and our fixation with heroes encourages us to believe that the single individual at the top is more singular than he really is. In several of the big companies we talked to . . . I had the impression that a dozen or so of their top people had the raw capacity to be CEO. . . . Their wealth of talent at the top, and manifest teamwork, makes for what many described as a seamless transition from one generation of management to the next. It is truly a team at the top that makes renewal happen.[4]

Similarly the successes in defense, medicine, and agriculture have not rested upon Einsteins, Henry Fords, Iacoccas, or Edward Tellers. Quite to the contrary, they have relied on what Galbraith referred to as "ordinary men" organized in an environment that allowed them to think new thoughts and to experiment.[5] Continuous innovation results from organizational arrangements that make thinking new thoughts and experimenting routine, everyday tasks.

The simple truth of the synthetic society is that the group is not only the source of ideas and the source of decision making, but it is also the vehicle for operation and management. The distinctive individual is not critical. Indeed, the Lone Ranger is a threat to success in the synthetic society.

The Free Market

The secular trinity posits that the largest number of individuals will benefit and society will gain the maximum benefits if the market is free. In the abstract, a free market is a place where there are many buyers and sellers and where buyers and sellers can enter, compete, and leave the market with ease. The United States does not have a free market, nor is one possible.[6] Governments always structure markets. Minimally government does that by enforcing contracts and protecting against fraud. In complex societies, government is pervasively and integrally involved in the market. In the United States it has been so for a long time.

American economic history is a history of market failures. The profession of economics exists largely for the purpose of trying to understand market failures. Most economists begin their investigations of market failures with the assumption that failures result from the market's freedom being interfered with in such a way that individual competitors can win without maximizing efficiency. In general, investigations of market failures seek to identify what factors cause market freedom to be interfered with. For the most part, the free market in the United States has been pragmatically defined as one not unduly impacted by pernicious forces. In the absence of agreement on a practical, real-world definition of a free market, it has been useful to define a free market as what is left when noxious influences are eliminated.

In earlier days, the primary noxious influences were monopolies and oligopolies. Monopolies and oligopolies exist when single buyers or sellers or small groups of buyers or sellers have the capability to dominate and manipulate the market and thus to win without being efficient. To protect against these noxious influences on the free market, the Congress passed the Sherman and Clayton Antitrust Acts and the Department of Justice established an antitrust division. Since World War II, the concern with monopolies and oligopolies has diminished considerably. By general agreement, the really noxious influence on the free market in the postwar period has been government. One need only monitor the political dialogue of the United States to know that government is commonly seen as the source of many of the nation's economic ills. This perception of the pernicious effect of government became ever more widely held as the nation's economic situation deteriorated rapidly during the 1970s and 1980s.

Although the dominant ideology characterizes government as a noxious influence on the free market, realism demands the recognition that government is necessary. Thus a great deal of effort in the post–World War II period has been devoted to defining the boundaries of acceptable government involvement in the market. Three kinds of government intervention have come to be recognized as either needed or acceptable.

First, there is a general consensus that government can and should act as a balance wheel to keep the economy from experiencing either excessive growth or excessive decline. Economists label this the "macro economic management role" and consider it necessary. Government exercises macro management by using its taxing and spending powers and its capability to control the money supply to constrain runaway growth and inflation or economic decline and unemployment. Under the rules that have been operative since World War II, the market remains free so long as the government intervenes only at the macro level. By contrast, micro management—that is, government intervention in specific sectors or product areas—is generally considered unacceptable. When micro intervention occurs, as it regularly does, it is attributed to inappropriate political pressure. Examples have been the Chrysler loan guarantee and a range of import restrictions. Such interventions represent a flawed system and are cause for lament. Indeed, it is part of the secular trinity faith that the market makes far better choices concerning whether to pursue ball bearings, bowling balls, ballet, or braunschweiger than government does. Thus there was no apparent reason for concern when services came to represent the overwhelming portion of the GNP and basic manufacturing activities went offshore. Such developments only reflected the market's pursuit of economic efficiency and that was to everyone's—but certainly to the United States's—benefit.

The second kind of acceptable government intervention occurs when important social or economic benefits or costs are not reflected in the private calculations of buyers and sellers. Thus it is acceptable for government to intervene in the market and disrupt normal efficiency calculations to protect Americans' health and safety. On this point there is a near consensus at the general level but continual controversy in specific applications. Throughout most of the post–World War II period there has been a significant increase in government intervention to protect health and safety. Beginning in the late 1970s and moving into the 1980s, there was something of a swing away from this. Most Americans, however, agree that this is an appropriate government role.

Finally, it is widely recognized that government must intervene in the market to purchase those goods and services necessary to carry out uniquely governmental functions. For example, government must buy those goods and services necessary to provide for national defense.

The free market has been interpreted to allow for the three types of government intervention. Nevertheless, when there is disagreement over how much intervention should occur, the dominant belief is that less intervention is preferable to more intervention.[7] All else being equal, American ideology prefers less government to more government.

The market, free or otherwise, is an abstraction. When government intervenes, it does so by influencing the decisions of buyers and sellers. In its role as macromanager of the economy, government seeks to manipulate excessive growth or recession by expanding or reducing the supply of money available to buyers and sellers. In periods of recession, government funnels more money into the economy and acts as a pump primer. In periods of expansion, it should—but seldom does—shrink available money. When government intervenes to protect health and safety, it creates conditions that make it advantageous for individuals to put more emphasis on health and safety than on efficiency. When government intervenes as a buyer, it becomes a competitor with other buyers and has the potential to create scarcity, thus driving up costs and reducing efficiency.

Even with the integration of the U.S. economy into a world economy and with growing evidence that the United States cannot compete, the secular trinity's answers remain the same on the subject of government's role. A common explanation for the United States's inability to compete in the international economy is that our competitors are playing unfairly; that is, they are playing according to rules that make it possible for them to win without being more efficient than we are. A frequent allegation is that the reason our competitors win is that in other countries there is cooperation between government and the private sector. The rhetoric of the debate is that all we want is "a level playing field" or the elimination of "unfair" competition. The assumption is that foreign governments tilt the playing field through their intervention in certain sectors of the economy in such a way that the United States loses. Presumably, the argument goes, with a level playing field the United States would win.[8]

The reality of the synthetic society is that the United States can no longer hope to succeed with a model of the free market that limits government intervention to the three traditional areas. As the United States has become increasingly interdependent with the rest of the world, its capacity to manage at the macro level has eroded steadily.

Pump priming in the United States, as is clearly indicated in the 1980s, is as likely to stimulate other countries' economies as our own, for with more money, Americans buy imports.

Another reality is that most of the world pays little more than rhetorical deference to the distinction between economic and political systems. The international marketplace is widely seen by other countries as an arena of national competition. In many countries cooperation among government, industry, labor, and financial institutions is not only acceptable, it is preferred. Similarly it is the norm throughout most of the world for governments to play a major role in the choice of areas, sectors, or product types in which the nation will be innovative and competitive.

The United States also is most competitive in those areas where government and the private sector cooperate in the innovation process—defense, medicine, and agriculture. This fact is illustrated in figure 4–1, which indicates the fifteen products that produced the largest trade surpluses and deficits in 1984.[9] Twelve of the fifteen products that delivered the largest trade surpluses are to a significant extent the results of innovations flowing from those three major sectors.

In the case of the five agricultural products with large trade surpluses, capturing the export market has been an explicit goal. Of the remaining ten large export earners, seven are spinoffs of the defense and medical systems. That is, they are commercially competitive products whose commercial export benefits were not a primary reason for their development. The economic sectors that perform poorly are those that lack synthetic organizational capabilities in the United States, and must compete with foreign producers who do have synthetic capabilities. This point is illustrated by the deficit end of the trade spectrum in figure 4–1. The fifteen products that accrued the largest trade deficits for the United States in 1984 are, with the exception of telecommunications and consumer electronics, produced by sectors without synthetic organizational complexes. In seven of them (fish and shellfish, consumer electronics, telecommunications, trucks and buses, nonmetal minerals manufacture, iron and steel, and passenger motor vehicles), foreign product and process innovations explain a significant and growing portion of the foreign advantage. At one extreme, it is not a lack of fish in American waters that explains our imports; it is the superior

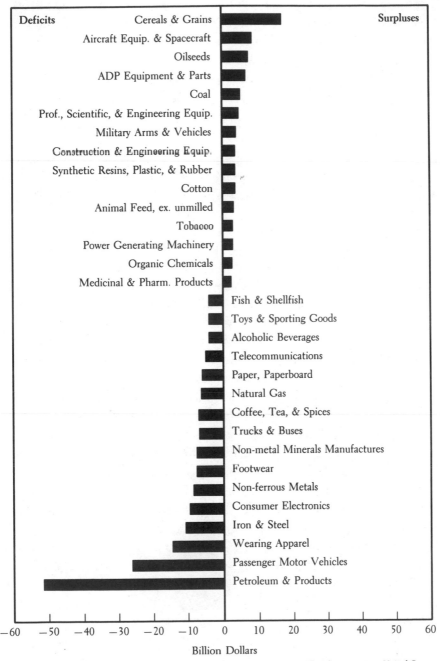

FIGURE 4-1
Largest Product Trade Balances, 1984

| Deficits | | Surpluses |

Cereals & Grains
Aircraft Equip. & Spacecraft
Oilseeds
ADP Equipment & Parts
Coal
Prof., Scientific, & Engineering Equip.
Military Arms & Vehicles
Construction & Engineering Equip.
Synthetic Resins, Plastic, & Rubber
Cotton
Animal Feed, ex. unmilled
Tobacco
Power Generating Machinery
Organic Chemicals
Medicinal & Pharm. Products

Fish & Shellfish
Toys & Sporting Goods
Alcoholic Beverages
Telecommunications
Paper, Paperboard
Natural Gas
Coffee, Tea, & Spices
Trucks & Buses
Non-metal Minerals Manufactures
Footwear
Non-ferrous Metals
Consumer Electronics
Iron & Steel
Wearing Apparel
Passenger Motor Vehicles
Petroleum & Products

−60 −50 −40 −30 −20 −10 0 10 20 30 40 50 60

Billion Dollars

SOURCE: Reprinted from U.S. Department of Commerce, International Trade Administration, *United States Trade: Performance in 1984 and Outlook* (Washington, D.C.: Government Printing Office, 1985), 13.

fishing and *processing* techniques and technologies of our competitors. At the other extreme—consumer electronics—it is the superior ability of our competitors to develop and rapidly produce higher-performance, higher-quality *products* at lower costs.

In 1987 the ten manufactured products that produced the largest trade surpluses, plus agricultural trade, and the seven product categories that produced the largest deficits suggest the same conclusion (see figure 4–2). The total trade surplus for the ten manufactured products plus agriculture was $35.2 billion.[10] Of the total, $25.6 billion, or 73 percent, was provided by aircraft, agricultural products, scientific instruments, military arms and vehicles, and medicinal and pharmaceutical products. All of these product categories are tightly linked to the defense, medicine, and agriculture sectors. In truth several of the other surplus product categories are probably beneficiaries of the synthetic organizational complexes in the same three sectors. The evidence is that where there is cooperation between the public and private sectors, the United States is competitive in the international marketplace. In defense and medicine, cooperation has been possible because of the ideological exclusions that allow government to intervene in the market for health purposes or to provide for defense. In the case of agriculture, the large government role is the result of historical accident and the political power of farmers that allows them to violate the trinity in the name of the family farm.

If the United States is to compete in the export race it will need the same kind of government/private sector cooperation that makes us competitive in the arms race. So long as Americans subscribe to the existing definition of the free market, commercially oriented public/private cooperation is difficult. As presently defined by the secular trinity, efficiency and competitiveness come with market freedom, defined as noninvolvement by government. The belief in this definition of the free market is more than just a barrier to competitiveness. It is a formula for failure.

FIGURE 4–2
Largest Manufacturing and Agricultural Product Trade Balances, 1987

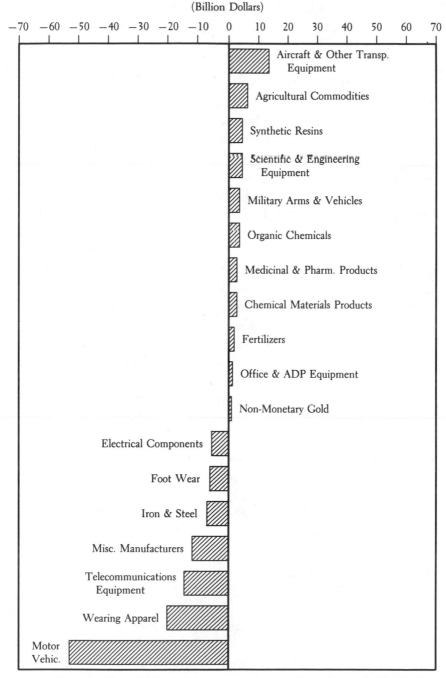

(Billion Dollars)

SOURCE: Adapted from data provided by U.S. Department of Commerce, International Trade Administration.

75

Efficiency

The secular trinity's mandate of efficiency as the operational goal for society is nearly universally accepted—government, universities, churches, in fact every organization in our society pursues efficiency. Frequently it is pursued even when no one knows how to define it, but central to the notion of efficiency is knowledge of how to do something. Efficiency is a comparative concept, a measure of the relative cost of producing goods or services. Efficiency may involve a comparison among producers, or it may involve a comparison between production at two different times.

In the synthetic society, however, larger and larger portions of production compete in a market where the standard of success is innovation, not efficiency. Recall that innovation involves the use of new ideas or methods and that it is carried out to accomplish one, some, or all of the following goals: superior performance, additional performance, new performance capabilities, superior quality, or efficiency (lower costs). In addition, in the synthetic society competition is structured by innovation-to-obsolescence cycles. Being the lowest-cost producer of an obsolete product or service has little value. Obsolescence is a driving concern. Thus the goal is increasingly to be first or, if not first, early in the production of goods and services that were not previously available. Being first provides a brief monopoly.[11] Being concurrent makes it possible to compete, but being late is to be not competitive at all.

Efficiency is not a standard that can be appropriately applied to the innovation process. How one measures efficient development and production of something that has never before been built remains an unanswered question. It is quite reasonable to make low-cost production a key goal in carrying out an innovation, but assessing the efficiency of the innovation process is not possible.

The broad utilization of efficiency as the operational goal of organizations has had serious adverse consequences for the United States. Perhaps the most serious is the tendency of American corporations, when they find themselves uncompetitive, to cut costs by cutting R&D expenditures and by eliminating experienced people. Given that innovation requires the ability to synthesize and that synthesis requires experiential organizational learning, laying off or firing personnel results in the loss of capabilities that are difficult to replace. Experiential organizational learning, as opposed to cognitive learning, is not something that can readily be hired in the form of a new specialist.

In the synthetic society, organizational networks and organizational complexes have the capability to do things no one person understands how to do. Success requires being continuously involved in an incremental experiential learning process. Breaking up the groups, teams, and organizations that have acquired incremental learning results in a serious loss of future capabilities, yet it is an instinctive response given the central importance our ideology assigns to efficiency.

The Secular Trinity and the Synthetic Reality

Thus an empirical look at where competitiveness occurs makes it clear that three of our most cherished ideas about economic activity have become counterproductive: groups, not individuals, are the source of success in the synthetic society; public/private sector cooperation, not the free market, is critical to successful competition in the international marketplace; and innovation—being first or early—is more important than efficiency in international competition. Our failure to recognize this new reality and our continued belief in the secular trinity has led us to misperceive how reality works and to prescribe inappropriate fixes for our problems.

PART II

ORGANIZATIONAL

SYSTEMS

WITH SYNTHETIC

CAPABILITIES

WORLD WAR II launched the organizational complex that would lead to the synthetic society. A series of ad hoc responses to wartime needs and opportunities mobilized science and technology in a way and on a scale never previously imagined. The innovation of weapons systems and military hardware during the war was nothing short of dazzling. Many believe the development or enhancement of such technologies as radar, sonar, acoustical torpedoes, and the proximity fuse, although less widely known than the atomic bomb, were in fact more important than the bomb to the successful prosecution of the war.

An examination of the organizational complex launched during World War II and of how it has evolved since offers critical insight into the elements necessary for continuous synthetic innovation in the United States. To build and sustain an organizational complex with the capability for continuous synthetic innovation, three things are necessary: first, a stable special-interest policy system is necessary to support the organizational complex; second, a rationale for violating the norms of the secular trinity must exist or be found; and finally, legal and administrative instruments must exist or be invented to provide for close, continuous, flexible interaction among organizations located in both the public and private sectors which have the diverse expertise needed for synthetic innovation.

Chapter 5 investigates the American policy-making system with respect to how special-interest policy systems are created or modified and how they make policy. A central objective of the analysis is to understand how the capability for continuous innovation has changed policy making. The size, political strength, and stability of the special-interest policy system for defense has provided the environment needed to achieve the remarkable record of synthetic innovation that has been attained in that sector over the last four decades. Only such special-interest policy systems can provide the sustained support (both policy and financial) needed to assure the continued successful functioning of synthetic organizational complexes. These policy systems are the support systems for the organizational complexes containing the organizational networks that deliver innovation.

Chapter 6 looks at how government, industry, and the universities were first linked during World War II. The war made it possible to disregard ideological prohibitions and overcome organizational inertia. Faced with the ultimate national threat, the nation did what worked— it invented the legal and administrative instruments necessary to accomplish what had never before been done.

The legal and administrative instruments necessary to link government, industry, and universities into an organizational complex capable of continuous synthetic innovation are described and assessed in chapters 7 and 8. At the core of the synthetic organizational complex for defense is the R&D contract, which gave government the ability to build and continuously modify the complex. The R&D contract allowed the defense agencies to build and maintain a set of public/private sector organizational linkages that together represent what Don

Organizational Systems with Synthetic Capabilities

Price, former dean of Harvard's John F. Kennedy School of Government, has called a contract federal system.

Although World War II successes, the national importance of defense, the power of its special-interest policy system, and the flexibility and creativity of the R&D contract have provided for continuous innovation in defense, the secular trinity is an ever-present disapproving presence. In practice, the defense complex has continually had to build and use ever more complex procedures to maintain the creative interaction that is necessary among the public- and private-sector organizations in order to produce continuous synthetic innovation. Chapter 9 provides a brief overview of the continual efforts aimed at forcing the defense complex to conform to the myth of the secular trinity.

At bottom, the ability to deliver continuous synthetic innovation is the same in medicine and agriculture as it is in defense. Nevertheless, the processes vary across these sectors in their fine details. To give a framework for comparison, chapter 10 provides a broad grid map of the origins, evolution, special-interest policy systems, and organizational complexes for medicine and agriculture. Most important, this overview makes it clear that the organizational complexes must be responsive to both the requirements of their substantive focus and the political needs of their environment.

Part II thus provides a description of American policy making, an in-depth look at the development and operation of the synthetic innovation capability in defense, and an overview of the synthetic capabilities in medicine and agriculture.

CHAPTER 5

The American
Policy-Making System

UNDERSTANDING the way public policy is made and executed in the United States is central to understanding how the nation created and has maintained a capability for continuous synthetic innovation in defense, medicine, and agriculture. It also explains, in part, the nation's inability to build a similar capability that is focused on the export market. Two characteristics of the American policy-making system stand out: it is fragmented, and it relies heavily on substantive expertise.

These two characteristics of the American policy-making system are reflected in our historical pattern of organizing policy around narrow sets of real-world activities and giving control of policy to those who have a special interest and expertise in the activities. The narrow substantive focus of policy is clear when it is recognized that, for example, "agricultural policy" is an umbrella label for what are really specific policies made for peanuts, corn, tobacco, and so forth. Similarly transportation policy is an aggregate of separate policies made for highways, railroads, airplanes, and so on. The historical tendency to organize policy around the narrowest possible substantive activity (e.g., corn or abortion) does not appear to be changing as the nation passes the second century of its Constitution.

For most narrow sets of substantive activities that are the focus of policy, there is a separate special-interest policy system that makes and manages policy. Some have suggested that giving policy control to special-interest policy systems explains much of the stability of the American political system.[1] By allowing those with a continuing vested interest in particular substantive activities (e.g., corn, highways) to make and implement policy for themselves, American policy making has escaped serious ideological, religious, and ethnic controversy. Thus inconsistency between a legislator's general position and individual votes is a given in our political system. For example, Americans find it perfectly understandable and acceptable that a solid conservative like Senator Jesse Helms of North Carolina, who is opposed to government subsidies, makes an exception in the case of tobacco. Similarly Senator Edward Kennedy, who appears to worry that military spending has gotten out of hand, is expected to support appropriations for New England's high-technology defense industry.

In organizing public policy around substantive activities and allowing those with a vested interest in the activities to make and manage policy, the nation responded to its physical circumstances. The system has served us well. For the first 100 years we had a seemingly endless geographical frontier; for the next 65 years the nation had a seemingly endless resource frontier; and for the last 35 years the United States has had an apparently endless technological frontier. Consequently, for most of its history the nation has not faced the need to redistribute wealth significantly. Instead, public policy has focused on expanding the use of abundant physical resources. American policy has sought to make both the rich and the poor richer by pushing back one or another physical frontier.

Whatever the problem or issue, the reflex response has been to use what Alvin Weinberg, former director of the Oak Ridge National Laboratory, has called the "technological fix."[2] In the nation's early years, free land served as both a relief valve for the discontented and a way to increase national wealth. At later stages, government land grants were used to develop railroads. Government built locks on its rivers and lakes, it built canals, and it provided generous leasing procedures for those who developed the mineral resources on the nation's public lands. Throughout American history a major response to public problems has been to use policy to accelerate economic and physical development.

Policy making in the United States takes place in two different arenas: the presidential/congressional arena and the special-interest arena.

Presidential/Congressional

The president and Congress, operating as a collective, address few policy issues. Presidential/congressional policy making generally deals with issues that generate broad national concern and have the prospect of creating political instability. Thus the president and Congress are directly involved in only a small number of the actions taken in and through government to resolve issues or solve problems. Policy made in the presidential/congressional arena is, however, the most difficult and important policy. Examples in the recent past have been environmental degradation, the energy crisis, and an unfair tax system. Presidential/congressional policy making normally requires value choices, is characterized by conflict, and frequently results in clear winners and losers.

Resolution of the nation's most difficult issues requires that presidential/congressional policy making either create or restructure special-interest policy systems so that responsibility for long-term management of particular physical activities can be handed to those systems. For example, when the Soviets launched Sputnik in 1957, space was immediately put on the presidential/congressional agenda, and over the course of the next few years, the president and Congress formulated a space policy. This involved creating NASA and giving it the charge of getting an American to the moon and back by the end of the 1960s.[3] In so doing, the president and Congress created a new special-interest policy system for space. Then, when the Arab oil boycott occurred in 1973, the president and Congress were faced with the task of formulating an energy policy. In that case they took control from five existing special-interest policy systems—those for oil, natural gas, coal, nuclear power, and electricity.[4] Subsequently, over the next seven years the president and Congress were involved in constructing a special-interest policy system for energy.

Before a topic on the presidential/congressional agenda can be handed to a special-interest policy system, four issues must be resolved: (1) the objectives to be pursued must be defined; (2) the boundaries of the physical activities that will be the focus of policy must be defined; (3) the instruments that government will use to manipulate the physical activities must be selected; and (4) the participants in the special-interest policy system have to be identified and provided with access to the policy-making process. Whether issues arrive on the presidential/congressional agenda for the first time, such as occurred in the case of space, or are moved forward from existing special-interest policy systems because of changed circumstances, as occurred in the case of energy, the president and Congress must find ways of resolving the four generic issues. When those resolutions are found, either a special-interest policy system is created or an existing system is modified. In either case, the special-interest policy system then assumes responsibility for the continuous formulation and implementation of policy.

Throughout American history, technological innovations have repeatedly caused the president and Congress to modify existing special-interest policy systems and create new ones. The innovation of the steamboat, the steam locomotive, the automobile, and the airplane all triggered demands for public-policy action and the establishment of special-interest policy systems. In the case of the steamboat, for instance, the frequency with which boiler explosions caused injury and loss of life led to the establishment of the first federal regulatory agency, the Steamboat Inspection Service.[5] The steam locomotive led to presidential/congressional policies that gave railroad companies large grants of land, which in turn provided the capital used to build many railroads. Later on, the natural monopolies represented by railroads led to demands for railroad regulation and the creation of the Interstate Commerce Commission. The automobile led to demands for a federal highway system, and the development of the airplane led first to the creation of the National Advisory Committee on Aeronautics, whose purpose was to push improvements in aircraft, and later to the large special-interest policy system associated with aircraft.

Prior to World War II, presidential/congressional policy making was primarily reactive to technology. The norm was for the president and Congress to establish special-interest systems when technological innovations generated serious national issues. World War II represented a turning point. In the years following the war, the president and Congress made major modifications in the special-interest policy systems for defense and medicine, the most important modification being the establishment of continuous innovation as a national policy goal. The combination of the Soviet threat and the World War II record of technological innovation resulted in the defense policy system being assigned, as a high-priority goal, the task of continuous weapons-system innovation. This, in turn, required redefining the boundaries of the defense system, expanding the range of policy instruments, and enlarging the range of participants in the defense policy community.

This special-interest policy system has supported and managed the development of the organizational complex that was largely responsible for moving the nation from an industrial to a synthetic reality. The policy-making system that has supported and managed defense innovations has operated with only limited public awareness.

Special-Interest Policy Systems

To appreciate what occurred, it is necessary to look at the historical structure and processes of special-interest policy systems and how they changed following World War II in those sectors that developed the capability for continuous synthetic innovation. What occurred in defense represented a fundamental change.

Special-interest policy making is primarily concerned with allocating benefits. Thus the actors involved operate in an environment where the struggle is over how much additional money, power, or jurisdiction they will enjoy. It has been a positive-sum game. In special-interest policy systems, policy making is concerned with a search for the optimum solution to problems, not with issue resolution (as presidential/congressional policy making is). Issue resolution requires choosing among competing values, while problem solving involves choosing the best option, given incomplete facts.

An underlying assumption of special-interest policy making is that, if complete information and understanding existed and all the concerned actors were fully informed, there would be no disagreement on the optimum policy. Central to the maintenance of this problem-solving environment is the role *experts* play in defining problems and proposing solutions. Experts on any set of substantive activities tend to share a common view of reality. Starting with that common view, they tend to define problems in scientific/technical/administrative terms, not in terms of values. Not surprisingly, the menu of solutions tends to be framed in similar value-free terms. Thus the way policy options are designed in special-interest systems tends to blur the value differences among actors.

The central role of experts in special-interest systems is reinforced by the tendency of nonexpert participants to defer to experts in defining the options. The popular notion is that experts define the policy options and others make the choices. A similar pattern of deference occurs among experts. It is felt that those with the most in-depth expertise should do the initial problem definition and option formulation. Thus the search for solutions to problems in special-interest policy systems tends toward ever more specialized categories. For example, defense policy consists of three major components: those focused on land, on the sea, and in the air. These are the responsibilities of subsystems built around the Army, Navy, and Air Force. Each subsystem is assumed to know its problems and solutions best. Within a given service subsystem, the pattern is for problem definition and policy option formulation to rest with more narrowly focused subdivisions—for example, infantry, artillery, armor, and so on. Again, the process relies heavily on narrowly specialized substantive expertise.

Appreciating how special-interest policy-making systems operate requires investigating three aspects of those systems: (1) the physical activities that are their organizing framework; (2) the individuals, interests, and organizations that participate in making and managing policy; and (3) the behavioral norms or decision rules that govern policy-

FIGURE 5-1
Special-Interest Policy Systems

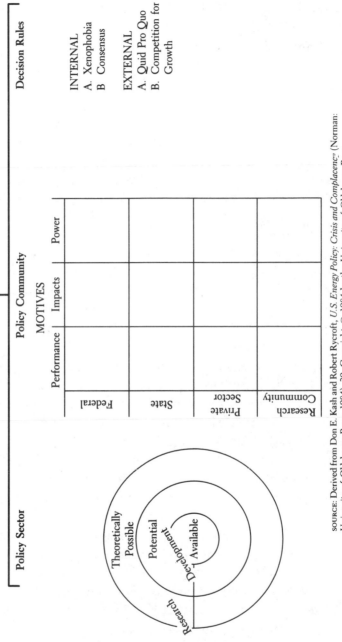

SOURCE: Derived from Don E. Kash and Robert Rycroft, *U.S. Energy Policy: Crisis and Complacency* (Norman: University of Oklahoma Press, 1984), 29. Copyright © 1984 by the University of Oklahoma Press.

making participants. In the following discussion we will refer to these as "the policy sector," "the policy community," and "the decision rules," respectively (see figure 5–1).

The policy sector denotes those substantive activities that are the focus, and thus the organizing framework, of special-interest policy. Policy is always preceded by a descriptor that refers to the sector. Descriptors such as agriculture, defense, medicine, highways, peanuts, or AIDS refer to distinguishable sets of physical activities or phenomena. Policy sectors are bounded by what it is physically possible to do or what participants in the policy system believe is physically possible.

The change in defense policy wrought by World War II was in the perception of what was physically possible—in sum, what was in the boundaries of the sector. Policy making had to deal with two new categories of substantive options: "potential" and "theoretically possible." Until World War II, the president and Congress organized special-interest policy systems around available activities. Available activities include those technologies, processes, and physical capabilities that are commonly referred to as "off-the-shelf." The distinguishing characteristic of available activities is that experts can provide reliable information on what they will cost and how long it will take to accomplish them. Thus decisions to buy or utilize state-of-the-art weapons technologies are informed by reliable information on cost and time. For example, before the war choices within the U.S. Army involved decisions about the appropriate mix of tanks, field artillery, or antiaircraft artillery, given finite money. In such cases the policy problem is deciding how to distribute reasonably predictable benefits among different interests. Throughout most of American history and in many special-interest policy systems today, decision making involves distributing known goods and services among competing interests. The organizational complex that was initiated during World War II and that innovated weapons systems created an expectation for continuous synthetic innovation in defense; thus it added the need to decide among potential and theoretically possible, as well as available, options. As the capability for innovation has grown over the last four decades, so have the policy options.

Potential activities are best exemplified by technologies that are at the development stage—for example, the next-generation aircraft. Such technologies require complex and high-cost technical development normally accomplished with prime contracts. Potential activities are distinguished by general agreement among experts that they can be developed and also by substantial disagreement about how long development will take and how much it will cost. The most frequent examples of cost overruns are associated with efforts to push potential activities into the "available" category. This is normally done by developing prototypes. Policy for available activities is concerned with the allocation of benefits when time and costs are known. Policy for potential activities must deal with the more difficult problem of allocating benefits when there is uncertainty about time and costs. Cornell University's Dorothy Nelkin, a recognized expert on technology policy, has noted that the existence of technical uncertainties "leaves considerable leeway for conflicting interpretations."[6] After World War II, the special-interest policy system for defense had to develop the ability to allocate resources among a growing number of potential technologies.

Theoretically possible activities are those that are possible within the laws of nature as we understand them, but that experts agree we don't yet know how to develop—for example, the hypersonic airplane. While there is agreement that these capabilities will be available at some point in the future, there is also agreement that the time and cost required to develop them cannot be reliably predicted. At this time the SDI, or at least versions and elements of it, probably falls somewhere in the boundary area between potential and theoretically possible activities. As the SDI illustrates, one policy option is to undertake R&D programs aimed at rapidly moving theoretically possible activities into the available category. When this occurs, the time and cost numbers used reflect acts of faith rather than figures derived from reliable calculations by experts. For example, advocates of the SDI program argue it will be cost-effective as a result of as-yet-unachieved technological breakthroughs.

Some things are not possible, even in those sectors with well-developed organizational complexes capable of continuous synthetic innovation. It is, for example, not possible to develop an antigravity machine. Such a machine would be inconsistent with our understanding of the laws of nature. At any point in time some activities are beyond the realm of the possible because they fly in the face of our scientific understanding of natural law; thus they are not policy options. Scientific understanding, however, is continuously advancing. Therefore, theoretically possible physical activities are continuously changing.

The range of policy options with which any special-interest policy system must deal is determined at the outside boundaries by our understanding of nature and, within those boundaries, by development of the expertise and organizational competence to carry out innovation. In some sectors—for example, railroads—the range of policy options has been, until recently, limited to available activities. Alternatively, in the medical sector the rapid development of biotechnology has led the medical research community to define theoretically possible options as first-priority solutions to many medical problems.

In defense, medicine, and agriculture, policy options are continuously being changed and expanded because the organizational complexes with synthetic capabilities are continually pushing both the boundaries of scientific understanding and the capabilities of technology. The direction and character of potential and theoretically possible activities is generally a reflection of R&D programs. Research money, whether spent for the sake of understanding (basic science), or because understanding is needed to accomplish something (applied science), and development money (funds devoted to creating prototypes) are the critical instruments used to change the boundaries of a policy sector. The R&D spectrum traverses the theoretically possible and potential options within the sector. Basic research begins at the outer boundary of the theoretically possible activities, with development (the delivery of prototypes) located at the boundary between available and potential activities (see figure 5–1). Since World War II, R&D in defense, medicine, and agriculture have steadily enlarged the policy options. The potential for absorbing both monetary and scientific/technical resources has expanded simultaneously.

In the 1950s and 1960s government was able to support almost all of the good basic science and a very large portion of the viable technologies proposed in defense. In the 1980s the menu of options is much larger than can be supported, and it continues to grow. The very success of defense and medicine in producing continuous innovation now poses serious allocation problems. To understand both the sources of success and the problems, it is necessary to look at the second component of special-interest policy systems—the communities that make and manage policy for those sectors.

THE POLICY COMMUNITY

The policy community consists of those actors who have a continuing interest in the activities included in a sector. They are organized to participate in policy making and have done so continuously. Participants in special-interest policy communities generally have three characteristics. First, they share a common view of the sector; that is, they agree on whether activities are available, potential, or theoretically possible. Second, they agree on who the members of their community are. And third, they share a commitment to the long-term promotion of the sector. Members of the community need not share common views on what policy should be pursued, however. Participation by actors who have not been continuously involved is resisted because they may not share a long-term commitment to the promotion of the sector and because they frequently are ignorant of substance.

Special-interest policy communities consist of organizations, interest groups, and individuals who are part of an information system primarily concerned with the generation, transmission, manipulation, and filtering of information about the substantive activities included in a sector.[7] Participants are normally located in four organizational settings: the federal government, state governments, the private sector, and non-profit research organizations (e.g., research universities) (see figure 5–1).

At the federal government level, for example, the defense community includes the following: defense committees in both the House and the Senate and appropriations subcommittees concerned with defense. Defense-related executive agencies include the Department of Defense, the Department of Energy, the Central Intelligence Agency, the National Security Agency, the Department of State, and the National

Aeronautics and Space Administration, plus a large and diverse collection of individual organizational components located in other federal departments and agencies.

At the level of state government, the National Guard is a continuing participant in defense policy making.

At the private-sector level, participants range from government contractors through a whole variety of organizations such as retired officers associations and the American Legion.

At the nonprofit research level, participants include the large community of university researchers whose work is funded by defense-related agencies or who serve as consultants to those agencies or to private-sector organizations linked to the defense agencies. Beyond the universities there are a large number of other nonprofit institutions linked into the defense policy system such as the RAND, Battelle, and Aerospace corporations.

The special-interest policy community for defense is further divided into numerous, more narrowly focused special-interest subcommunities—those organized around the Army, Navy, and Air Force, for example. Within the service subcommunities are more specialized communities built around the organizational networks that deliver and use particular weapons systems. Few of these participants look at defense policy in broad strategic terms. Rather, most participants are involved in making and implementing policy for particular weapons systems or subsystems.

To appreciate why participation occurs, it is useful to look at three motivations for participation: performance, impact, and power.[8]

The Performance Motive

In those systems with the capability for continuous synthetic innovation, the performance motive deserves special attention. Performance has to do with improving understanding and/or enhancing or creating physical capabilities within a sector. In those sectors with capabilities for continuous synthetic innovation, the pursuit of superior performance is a continuous goal. Two expectations underlie this drive. One is the general tendency of Americans to seek technological fixes to public problems. Thus performance is pushed in a search for those fixes. The other is that continuous pursuit of superior performance will open new opportunities and/or provide the basis for solving future problems.

The American Policy-Making System

Scientists and engineers frequently participate in policy communities predominantly for performance reasons. They participate primarily by working in their professional capacities, whereby they develop the knowledge and capabilities needed to define public problems and formulate policy options. Thus, by pursuing their professional goal of superior performance, experts participate in policy making.

To be an expert in the United States normally requires training at a university level, minimally a bachelor's degree, and, where the role is research and development, a Ph.D. Thus policy options that involve technological fixes are largely the products of people who have been trained and socialized to be researchers. A central part of that socialization involves instilling the belief that increasing knowledge or increasing technological capability is an a priori good. Thus, in addition to strong societal inclinations toward technological fixes, there are, distributed throughout special-interest policy communities, networks of experts socialized to believe in and take as a primary mission pushing the state of the art. Since World War II, scientists and engineers have become increasingly common in every one of the four organizational settings of the defense special-interest policy community. A similar pattern exists in medicine and agriculture. Congressional staffs and executive agencies associated with defense are now permeated by people with technical training. A growing number of military officers have graduate training and sometimes Ph.D.'s. The defense industry and, of course, the nonprofit sector are permeated by scientists, engineers, and technicians.

The importance of the performance motivation in special-interest policy making is reinforced by the fact that other participants are inclined to accept the consensus view of experts. A consensus view by scientists and engineers carries great influence both because of their expertise and because they are frequently seen as valuing scientific understanding and technical performance over the less virtuous pursuit of profit or power. In special-interest policy communities, scientific/technical participants define which activities are available, potential, or theoretically possible; and those definitions are, for the most part, relatively noncontroversial. Scientific/technical members of special-interest policy systems, regardless of their organizational location, normally enjoy a shared perspective founded on a common body of theory, data, and experience. They frequently share what philosopher of science Thomas Kuhn has called a "paradigm."[9] The powerful need for consensus in special-interest policy systems gives great influence to those who can provide it.

The Impact Motive

Participation driven by impact, the second motive, is commonly recognized. Impact-motivated participation reflects a concern for how physical activities affect phenomena that participants value. The most common impact motivation is economics. Thus a major motive for aerospace companies' participating in the defense-policy community is concern for profit. Participation by the Department of Defense is driven by a concern with development of those technologies that provide the best protection against foreign military threats. Some continuing players in special-interest policy systems participate because they are concerned with the impact of activities on the environment or other valued phenomena. In sum, the impact motivation refers to a concern with how activities in a sector affect the surrounding environment.

The Power Motive

Power is the third motive for participation in special-interest policy communities. The power motivation reflects a desire to protect, establish, or expand influence over the physical activities. Concern with power is frequently identified with government participants who are seen as placing desire for influence or control before concern with performance or impacts.

Government agencies normally have groups of managers who are continually and actively involved in protecting their organizations. Some suggest that concern with organizational well-being and growth results from the desire of civil servants to protect their jobs, but this explanation seems too narrow. Participants in large organizations often tend to identify personally with the organization, so that its growth and well-being become inseparable from the individual's self-image. Out of this interaction appears to come the powerful desire of organizations to protect and expand their turf.[10] In truth, the pursuit of power is evident in most organizations and interest groups, whether located in the public or private sectors.

Few individuals, organizations, or interests are driven by a single motive. In the real world, actors in special-interest policy systems participate for a mixture of motives. Whatever the motive, an effective role in special-interest policy communities requires continuous participation, which in turn requires organization. Organization is necessary, in part, to stay abreast of complex scientific/technological changes. Policy communities are informal systems that link Washington (or state capitals) with other organizational components in society, and those components communicate with each other constantly. They are information networks that deal primarily with substantive information and operate with little regard for organizational or hierarchical positions.

The way in which information is processed, communicated, filtered, and analyzed in defense is illustrative. It ranges from the telephone to a floating cadre of people who fill a significant portion of the seats on many commercial airplanes worldwide. It is a system served by sophisticated audio and visual communication systems as well as by hard copy. The more important point, however, is not the vehicles of communication; it is that meaningful participation requires substantive information on and understanding of the physical activities. Each special-interest policy community is a unique technical-language network. To understand what is going on, one must understand the shorthand, and that requires state-of-the-art knowledge. To enter the system and participate seriously requires a major effort. Alternatively, lack of substantive knowledge is an effective justification for special-interest communities that exclude those who try to participate without appropriate organization.

DECISION RULES

Critical to understanding special-interest policy making and implementation is an appreciation of the rules governing the decision process. It is a process in which government is best seen as an issuer and implementor of policy evolved through accommodation among the participants located in each of the organizational settings. Two sets of decision rules are essential to the successful functioning of the multiple special-interest policy systems. One set (internal rules) governs the interaction of those who participate in the individual communities; the other (external rules) governs the interrelationships among the different special-interest policy communities.

Internal Rules

The internal rules are xenophobia and consensus. Xenophobia refers to the shared obligation of community participants to exclude nonmembers, i.e., any actor who has not been a continuing participant. Central to the xenophobia rule is a prohibition against any participant's going outside the community to seek allies to achieve a narrow or specific policy objective. The xenophobia rule says that the stability of the community is more important than any participant's single goal.

The resistance to outsiders reflected in the xenophobia rule results from two concerns: outsiders cannot be expected to share the commitment to protect existing benefits. Recall that special-interest policy systems are involved with allocating benefits. Policy struggles concern the relative distribution of these benefits. Therefore, participants always start with a need to protect what they have. New participants are not likely to have been beneficiaries and may actually have the redistribution of benefits as a goal.

Outsiders also pose a threat because of substantive ignorance. Specifically, they may not share the same view of what activities are available, potential, and theoretically possible. Especially in sectors such as defense, where a capacity for continuous innovation exists, technology has become exceedingly complex. Only continuing participants understand the potentially damaging or disruptive consequences of what may appear to the ignorant to be minor substantive changes. Modern defense technologies are systems composed of interdependent subsystems and components. Changes in performance requirements, components, or subsystems frequently require changes in other parts of the system, thus significantly altering their performance. Fear of ignorance and fear of a lack of commitment to the existing distribution of benefits represent powerful enforcers of the xenophobia rule.

The consensus rule goes hand in glove with xenophobia. Consensus means that no changes in policy will occur until a consensus exists. Consensus does not imply that every participant agrees; rather, it means that no change in policy will be undertaken that causes participants to organize to oppose it. Decision making by consensus gives each participant veto power, but it requires that a veto be exercised only when participants see policy changes as fundamental threats. In sum, the solidarity and stability of the community take precedence in all but exceptional circumstances. Hans Landsberg of Resources for the Future has captured what consensus means when he says it requires "a willingness to moderate demands for adopting one's entire agenda."[11]

The ability to achieve consensus in special-interest systems relies heavily on the role substantive experts play in defining options. As previously discussed, experts start with the assumption that, if full understanding were available and everyone had it, there would be an optimum policy choice. The central role of experts in defining options and as gatekeepers in the communications networks of special-interest policy systems results in a powerful tendency to blur value differences, which reduces the likelihood of polarization. Within special-interest policy systems, then, debate over policy options and the evolution of policy occurs in a context where competing participants share a common set of objectives; a common view of what is physically or technically possible; a belief that the most successful route to the objectives involves the right choice among physical or technical options; and the assumption that, if total information and understanding were available, there would be one right solution. In combination, these assumptions, along with the fear that outsiders will be disruptive, make consensus decision making an imperative. Given the xenophobia and consensus rules, policy within special-interest policy systems must necessarily proceed in an incremental way.

External Rules

For policy making to evolve as just described also requires that special-interest policy communities not interfere with and disrupt each other. American policy making has reflected a powerful characteristic: those with a vested interest in a particular sector can make policy for themselves. To protect the jurisdiction of special-interest policy communities has required evolving rules that govern their interactions with each other. They are the "quid pro quo" and "competition-for-growth" rules.

The quid pro quo rule says, "We will not interfere with your right to make and implement policy for your sector if you do not interfere with our right to the same for our sector. Should you interfere, however, we will reciprocate." The quid pro quo rule involves a commitment by special-interest policy communities never to take actions aimed at *explicitly* appropriating rights or benefits already held by another special-interest policy system. These rights and benefits range from governmental revenues to organizational and legal jurisdiction. In sum, the quid pro quo rule says, "Anything you have is yours. Should any other special-interest policy system try to take it away, you have the right to attack its vested benefits and try to take them away." This rule is eroding, however, as the budget deficit undercuts the workability of the rule described next.

The competition-for-growth rule responds to the need of special-interest policy systems to grow. The quid pro quo rule prohibits expansion by means of expropriating anything that another special-interest policy system has. The competition-for-growth rule says that you have total freedom to compete for and try to capture as large a portion of uncommitted resources or areas of jurisdiction as you can. The competition-for-growth rule rests upon the American experience of ever-growing abundance. It is the relief valve for a policy system heavily comprised of special-interest policy systems, all seeking to expand their turf. The American experience has been one in which the pie has grown steadily. Thus Americans assume that there will be new, uncommitted resources available for distribution year after year. Recent budgetary austerity has diminished but not destroyed this fundamental expectation. The assumption of endless growth applies to organizational activities and legal jurisdictions as well as to economic and physical resources. The competition-for-growth rule says it is acceptable to compete for that portion of the economic pie or those new areas of jurisdiction that are not already committed, and it assumes that growth will itself provide uncommitted resources and activities. The assumption is that no one has to give anything up to permit others to improve their position. The fact that this has occurred historically means that special-interest policy systems operate in an environment where policy has involved the allocation of additional benefits, not the reallocation of finite benefits.

Increasingly, over the post–World War II period, the competition-for-growth rule has become integrally linked to synthetic innovation. As a result, special-interest policy systems and their subsystems see gaining jurisdiction over new technologies as a way to ensure a larger cut of annual growth. Within the Department of Defense, one of the competitive mechanisms of the services has been to seek jurisdiction over radical innovations. This general phenomenon was illustrated in the energy sector in the days immediately following the Arab oil embargo, when a range of special-interest policy systems sought to gain jurisdiction over various energy technologies. Certainly one stimulus was the expectation that jurisdiction over technologies would result in larger budget allocations and larger areas of policy jurisdiction.

The perceived link between technological innovation and competition-for-growth is illustrated in a different form by defenders of defense research and development. R&D, it is argued, not only will contribute to the nation's security, but will actually generate multiple commercial spinoffs. For example, General James Abrahamson, Jr., the first director of SDIO, argues that commercial spinoffs will pay for the SDI.[12] Broadly speaking, then, innovation is considered not only to provide a powerful vehicle for capturing a share of growth; it is also expected to generate that growth. Thus military innovation is characterized by some as the perfect "positive-sum game."

In the end, participants in special-interest policy systems behave as they do because they know that if either the internal or the external rules are violated, instability will result. When instability occurs, demands develop for the president and Congress to take jurisdiction of the policy sector. When the president and Congress take jurisdiction, the special-interest policy communities lose control and exist in that most unpleasant of worlds, one in which policy outcomes are unpredictable.

Current Stresses

In the late 1980s the American policy system is increasingly stressed. Contrary to the expectation that has traditionally underpinned the competition-for-growth rule, the United States is not experiencing sufficient growth in the production of manufactured goods to deal with the demands for additional benefits by the special-interest policy systems.

This dilemma is particularly evident in defense and medicine. Having built organizational complexes with the capability for continuous innovation, these systems are creating more and more complex technologies. At the same time, they are more costly in both dollars and scientific/technical resources. Recall that special-interest policy systems only have the capability to allocate benefits. When costs must be allocated, presidential/congressional policy making is normally required. Thus defense and medicine require more resources because of continuous innovation. At the same time, the U.S. economy has been integrated into the international economy and is becoming less competitive, so that it provides fewer resources to allocate. The short-term fix has been to borrow heavily from foreign nations.

Within the defense special-interest policy system the nation has been continually recommitting itself to expensive new weapons systems, while the resources available to produce them have been declining in a relative sense. Since only the most advanced technological options are believed to offer security, the defense special-interest policy system continues to push the state of the art. Two things are occurring. First, the time required to bring new innovations to prototype is increasing.[13] Second, the armed services are buying smaller and smaller numbers of weapons systems. The problem is illustrated by an apocryphal story told within the U.S. Air Force, in which it is said that at some point in the future the Strategic Air Command will have three hypersonic airplanes—because that is all they will be able to afford. One will be stationed in Asia, one in Europe, and one in the United States, and none of them will be flown because the Air Force will not want to risk having an accident with one of them. The situation would be analogous to the Navy's apparent unwillingness to use aircraft carriers in the Persian Gulf during the Iran-Iraq war.

The ever-expanding special-interest policy system demands that are driven by continuous innovation, combined with a lack of growth in the economy's ability to support these demands, suggests that the American policy system is traveling on a very dangerous path.[14] There are only two options. One is to cut back on the benefits distributed by the special-interest policy systems. This is obviously occurring already in many special-interest policy systems. Note, however, that it has not been occurring in defense, medicine, and agriculture, all of which have enjoyed during the 1980s an increasing share of society's resources, certainly of government resources. The other option is to strengthen our capability for producing manufactured goods that are competitive in the international market. That capability presumably would allow the United States to experience growth in its productivity and thus pay for the continuance of the present policy-making system.

CHAPTER 6

The Development of
the Synthetic Capability

THE organizational complexes that produce continuous innovation in defense, medicine, and agriculture all function in ways that violate the standards of the secular trinity. These complexes are characterized by intimate and continuous cooperation among government, industry, and universities. In each of the three sectors the circumstances that allowed the strictures of the secular trinity to be violated have been different.

Because of ad hoc responses to the need for innovating weapons systems and military hardware, World War II was to be the trigger for the revolution reflected in the synthetic society. Around each distinctive innovation of military technology an organizational network was assembled for the purpose of pushing the state of the art. By the end of the war, these networks together represented a new capability, an organizational complex with the capability for continuous synthetic innovation.

Vannevar Bush was head of the Office of Scientific Research and Development (OSRD), created by President Franklin Roosevelt to mobilize the nation's scientific and technical resources. As a key architect of the new organizational complex, Bush appreciated the revolutionary character of what had been invented. He saw that the real

The Development of the Synthetic Capability

genius of the system lay in the ability to link government, industry, and university researchers together. These three previously separate repositories of expertise and skills, when joined, produced synergistic capabilities.

Bush described what occurred as follows:

> On the one hand were military men, burdened with the extreme responsibility that only military men can carry off ordering great numbers of their fellows into strife where many of them must die, harassed by an unreasonable load of work, determined to get on with a tough and appalling job and get it over. On the other hand were the scientists and engineers, realizing from their background that much they saw was obsolete, forced to learn overnight a new and strange way of life and set of human relations, driven to the limit by the keen realization of the scientific competence of the enemy and the consequent desperate nature of the race. And in between were industrialists, faced with the production of unprecedented quantities of strange devices, wary of abrupt changes that would wreck the mass production in which this country excelled. These diverse groups and points of view could not collaborate unless they were forced to do so artificially and ineffectively by arbitrary orders from above, or unless they learned a new partnership. They accomplished the latter, and the accomplishment was greater than the mere creation of new weapons, and made such creation possible on a scale which determined the outcome. Out of it evolved, toward the end, an effective professional partnership of scientists, engineers, industrialists, and military, such as was never seen before, which exemplified the spirit of America in action at its strongest and best.
>
> This is the story of the development of weapons of war, but it is also the story of an advance in the whole complex of human relations in a free society, and the latter is of the greater significance.[1]

The development of radar illustrates both the launching and the style of operation of the organizational networks that were rapidly created during World War II. At the time hostilities began in September 1939, the United States, Great Britain, France, and Germany were all working secretly and independently on radar. The great breakthrough in radar that was to be the basis for the striking developments in the United States was provided by the British. It was the resonant cavity magnetron, which remains the heart of all modern radar systems.

The critical event was the creation of the Radiation Laboratory at the Massachusetts Institute of Technology on 10 November 1940.[2] This government-financed, university-based laboratory coordinated efforts of university professors, private industry, and government for research and development. The Radiation Laboratory, in collaboration with five large companies that had contracts to produce components, made rapid progress. By 4 January 1941 an operating radar was working on the roof of the main MIT building. Out of the Radiation Laboratory were to come some 150 different radars that can be roughly divided into three categories: ground radars, ship radars, and airborne radars. Their contribution to the war effort can hardly be exaggerated.

The challenges posed by the need to build reliable, precise, powerful radars to meet a wide range of needs were immense. Some sense of the pattern of the development can be seen by looking at one of the first major problems—the development of a magnetron with substantially greater power than the British model. A large number of scientists and engineers were asked to move to Cambridge to work on radar. Science reporter Daniel Greenberg summarizes the memories of Nobel laureate physicists I. I. Rabi and Luis Alvarez as follows:

"We worked from the British experience," said I. I. Rabi, who served as associate director of the Rad Lab. It was obvious at the outset, he explained, that nobody knew anything about radar, but the physicists had "the intellectual mobility" to find out. "They were light on their feet," says Luis Alvarez, who served at the Rad Lab and in the Manhattan Project. "They knew something about electronics because of their work with accelerators, but the real reason was that they were the best people and they were adaptable to anything." Rabi adds, "The people at the Rad Lab didn't know a damn thing about radar. I was in charge of a magnetron group and I'd never seen one. Well, I thought, I'll go around to MIT and ask some of the electrical engineers. After talking to them, I could see that they didn't know anything either. . . . So we started absolutely fresh and designed magnetrons." An immodest, arrogant assessment of the wartime experience? Perhaps. But the physicists were "light on their feet," for in short order they achieved stupendous success in work distantly related to their peacetime occupations.[3]

Thus, faced with not knowing how to do the job, Rabi and his colleagues used the body of experience and theory that they had developed through years of basic research to design a more powerful magnetron. In numerous other cases throughout the war effort, comparable organizations that allowed scientists with theoretical knowledge to be mobilized for the purpose of pushing the technological state of the art were established.

Contributing to the war effort involved more than design and development, however. It involved producing and deploying new technologies rapidly and in large numbers. Government had to link university researchers with the industrial organizations that would manufacture the technology and with the military organizations that would use it. An organizational network with the capability for continuous synthetic innovation of radars was built on an ad hoc basis. It was built because accomplishing the needed innovation required mobilizing the range of people perceived as having some chance of contributing to the innovation, wherever they might be located, and linking them together in a network that could effectively develop, produce, and use the technologies.

The striking discovery of the war, then, was what could be accomplished when basic scientists, industrial organizations, and military users were linked under conditions of organizational flexibility and compelling national need.[4] It must be emphasized that what occurred in the cases of radar, sonar, the proximity fuse, and so forth, was not the implementation of a grand design; rather, these successes represented pragmatic responses to specific needs. There was one governing rule: do what needs to be done to accomplish the task.

Wartime scientific/technological developments were possible because immediate military needs were perceived as being more important than regulations, organizational rules, or even broad philosophical and ideological standards such as the secular trinity. Thus intimate cooperation among government, industry, and universities was acceptable during the war, whereas before it had been unacceptable.[5] Crisis allowed the traditional barriers in society to be crossed and the traditional bureaucratic and legal rules to be changed.

Before the War

To appreciate the revolutionary change in relationships among government, universities, and industry, it is useful to look at each institution's use of and support for science and technology prior to the war. First, each institution was seen as having a different and distinctive mission. Second, each institution sought to be self-sufficient in the science and technology needed to carry out its mission. Third, each institution saw intimate relationships with the others as having little value and some potential danger.

MISSION ORIENTATION

Before the war federal agencies saw science and technology as instruments to be used to accomplish clearly defined agency missions. Federally funded science and technology had to be justified and understood in terms of specific programs.[6] If the mission value of scientific technical activities could not be explained to and understood by laymen, the activities would not be supported.

Most prewar federal scientific/technical programs were contained in four distinct and bounded areas—agriculture, defense, aviation, and natural resources—and the bodies of expertise used were relatively self-contained and applied. Prior to the war, roughly 40 percent of federal expenditures for science and technology went to agriculture, the agricultural sciences being divided into such categories as animal husbandry, agronomy, and the plant sciences.[7] Agricultural research and knowledge, then, were organized around economically important plant

The Development of the Synthetic Capability

or animal species or around such practical problems as soil fertility or erosion. By comparison, in the traditional scientific community, work was divided by disciplines such as chemistry, physics, biology, and so forth.

In defense, research and development was organized around particular technologies, such as tanks, ships, and airplanes, and was conducted in arsenals and shipyards. The scientists and engineers involved had applied orientations and skills, as opposed to theoretical orientations. For example, the work on airplanes funded by the National Advisory Committee on Aeronautics was generally carried out by engineers and relied heavily on empirical experimentation. The same was true in the natural resource area, where the geologists (heavily concentrated in the U.S. Geological Survey) tended to specialize in coal, oil, or particular metals or, alternatively, in mapping geologic structures defined by quadrangles on maps.

Federally funded science was mostly descriptive and empirical in character and was little concerned with the development of theoretical structures. The idea of government supporting work aimed at the development of theory with no clearly identified utility was foreign to the system. Harvard's Harvey Brooks describes an example:

> [A]lthough agricultural science did make important use of genetic principles and contributed toward their development, the support of genetics as such because of its importance to agriculture was not generally recognized as a federal responsibility. If one could gain an understanding of genetics through the study of an economically important plant or animal species, this was all to the good. But it was not thought legitimate to support the study of genetics using an economically worthless organism just because such an organism might be suitable to achieving rapid advances in theoretical understanding.[8]

AGENCY SELF-SUFFICIENCY

With the exception of agriculture, where a large portion of the research was carried out in land-grant colleges using federal funds transferred to the states on a formula basis, government scientific technical work occurred in-house. That is, it was carried out in government laboratories or facilities by scientists and engineers who were members of the Civil Service.[9] The idea of contracting out for science and technology was not even an option that had been considered. The relatively narrow applied focus of government R&D and the limited and self-contained kind of expertise required to accomplish the R&D tasks of the agencies made it relatively easy for government agencies to be scientifically and technically self-sufficient.

In the case of the military services, there was another and even more compelling reason for being technically self-sufficient. It rested on the belief that government procurement of weapons from industry required an arms-length relationship.[10] To accomplish this the Army operated a system of arsenals and the Navy a system of navy yards, both having the capability to design new hardware and produce limited quantities of it. The goal was to have in-house capability to perform all of the technical functions associated with developing and maintaining weapons systems. The rationale was that civil service teams of engineering and production personnel gave the agencies a capacity to evaluate and manage contracts with the private sector and provided government decision makers with disinterested advice.[11] The ideal arrangement was one where civil service employees designed and produced limited quantities of new weapons systems that could, when needed, be turned over to industry for quantity production. This provided the government with what came to be known as a "yardstick" by which government could measure the efficiency and performance of private-sector producers.

LIMITED INTERACTION AMONG INSTITUTIONS

In American industry the attitudes toward and the use of science and technology had many parallels with those of government. Prior to the war most industrial companies were organized to produce relatively distinctive product lines such as automobiles, steel, rubber, chemicals, and electrical equipment. The nature of these products did not change very rapidly, and the operational goal of American industry was increased efficiency in production. For the most part, each company sought to develop and utilize its own in-house scientific technical capabilities as was deemed necessary to maintain competitiveness and profitability. As in government, the scientific technical skills used in American industry were provided mostly by people with applied kinds of expertise, particularly engineers. With the exceptions of the chemical industry and to some extent the electrical industry, American companies saw little need for or utility in gaining access to the basic sciences. There was little need or desire for close cooperation with universities or government. American industry generally saw government as either a buyer of its products or a regulator. In general, then, industry shared with government a preference for an arms-length relationship.

Understanding the relationship among universities and government and industry prior to World War II requires distinguishing between the land-grant institutions—more precisely, the colleges of agriculture—and the rest of the university system. The agriculture schools received substantial federal support and had applied missions. The rest of the university system had little or no government support and a limited applied mission.[12] American universities generally were seen by themselves and others as institutions primarily involved in providing education and carrying out basic research—that is, research aimed at expanding knowledge regardless of its social utility. The primary "audience" of university researchers was other researchers, located for the most part in other universities. The highest-priority goal was advancing knowledge, ideally by developing new theoretical structures.

Government did not support university research for two reasons. First, it was not seen as making a direct contribution to governmental missions. Second, it was widely agreed that the federal government had no role to play and, in fact, should not be involved in support of education. Support for education and research was generally perceived to be the responsibility of state governments and private donors. Much the same set of attitudes characterized the approach of American industry to the universities. For the most part, when industry supported universities it was through industrial foundations committed to general educational and research support and not with the idea of funding research that would be of direct value to industry.

From the university side, there was a continuing fear that federal support, whether for education or research, might infringe on and misdirect the university's activities.[13] One concern was that with federal money might come control that would limit the right of the universities to investigate and debate ideas freely. Researchers feared that with federal support would come efforts to focus the direction of research. In fact, the example of the land-grant institutions and the agricultural colleges illustrated the danger. Specifically, their need to focus on economically or socially useful tasks limited their ability to focus on the most intellectually exciting frontiers. Thus the university community had little more interest in receiving federal support for research than the government had in offering it.

The World War II Transition

Harvey Brooks has accurately emphasized that the years between 1941 and 1945 "represented a remarkable watershed" in the relationships among government, universities, and industry.[14] In meeting its wartime needs, the government built organizational networks to innovate technologies that linked basic research, applied research, development, production, and deployment into dynamic, synergistic systems. To accomplish this it was necessary to shatter the separations and barriers to interaction that had historically characterized the three institutions. Arms-length relationships were converted into intimate cooperative cross-fertilizing relationships. The systems both facilitated the rapid transfer of scientific theory into applications and stimulated a rapid acceleration in research and technological capability. Military liaison people were posted at research laboratories in American universities. Cooperative research and development was carried on by teams made up of university scientists, government researchers, and industry personnel. University scientists not only participated in conceptualizing and developing prototype technologies but also participated in field tests, in the development of strategies and tactics for using the technologies, and in training military personnel how to use them. A set of institutions that had previously been divided by walls of separation became virtually a seamless web.[15]

112

The Development of the Synthetic Capability

The flexibility and adaptability of both government and industry were remarkable. But, by general agreement, the greatest surprise of the World War II experience involved the degree to which theoretically oriented basic scientists were able to mobilize and contribute to the war effort. It was in the striking extent to which basic scientists were tapped to do so that World War II changed things for all time. The atom bomb development most graphically symbolizes what occurred. Richard Rhodes, Pulitzer Prize–winning author of *The Making of the Atomic Bomb,* describes in the opening of his book one of the ways in which development began.[16] A *New York Times* review summarized that opening:

> Mr. Rhodes' book opens in 1933 on a street corner in London, where the physicist, Leo Szilard, 35 years old, a Hungarian-born Jewish refugee from Nazi Germany, pondered a remark in that morning's "Times" that said talk of unlocking the atom's hidden power was so much moonshine. "Szilard stepped off the curb," Mr. Rhodes writes. "As he crossed the street time cracked open before him and he saw a way to the future." Szilard's vision was that subatomic particles known as neutrons could slip through the electric barriers that held that atom together, breaking it apart and triggering a chain reaction that would liberate nuclear energy for industry or war. Knowing no specific way to start a chain reaction, Szilard was nonetheless so confident it could be done that he filed a patent on the idea and eventually became a driving force behind the Manhattan Project.[17]

The lore about the role of scientists in World War II is replete with similar stories. James Phinney Baxter, in *Scientists Against Time,* the official history of the Office of Scientific Research and Development (OSRD), documents the breadth of technological breakthroughs and advancements and the role of university scientists in those developments.[18] The grafting of scientific knowledge and the skills, habits, and attitudes of the scientific community onto industry and government opened up possibilities previously imagined only by science-fiction writers.

What was it that allowed this grafting to occur? And, more specifically, why was the scientific community able to make contributions that it had never made before? Two factors offer much of the explanation for the ability of the university-based science system to rapidly respond to the needs of war. The first was the availability of a cadre of truly remarkable scientists. The second was the invention of a new way for government to mobilize and utilize university-based talent.

HIGHLY COMPETENT SCIENTISTS

To appreciate what occurred, it must be remembered that prior to the years between World War I and World War II, with rare exceptions the American scientific community was not distinguished. Frontier science was primarily the province of European researchers. Beginning in the 1920s, however, there appeared in the United States in certain areas scientists of truly international renown.[19] By the late 1930s, Americans were making important contributions to pure theory in physics, medicine, and astronomy and establishing themselves among the world's scientific elite. This was especially true in atomic physics.

As the quality of American science improved, American scientists became increasingly linked with the European scientific community. This linkage was to have high payoff when, during the 1930s, many of Europe's leading scientists moved to the United States. The migration was driven by the rise of totalitarian dictatorships in Europe and occurred in an ad hoc fashion. American scientists found positions in their universities for European colleagues seeking to escape fascism.[20] No other nation ever enjoyed the addition of so much scientific talent and quality over such a short period of time.

In combination, then, a maturing basic science community and the addition of emigré scientists from Europe provided the nation with a pool of talent that was to be mobilized on an unprecedented scale and with startling speed to support the war effort. At least three factors contributed to the speed and enthusiasm with which the scientific community threw itself into the war effort.[21]

First, patriotism played an important role. Like other Americans, scientists were eager to make the maximum possible contribution to

The Development of the Synthetic Capability

winning a war that clearly pitted good against evil. The Japanese attack on Pearl Harbor left no doubt as to who the aggressors were.

Second, fascism was a totalitarian ideology. It was predicated upon an idea system that demanded that people not only behave in ways consistent with the ideology but also that people think correctly. Thus the ideologies of our nation's enemies were perceived as being uniquely threatening to the very essence of science—threatening to its need for the free exchange of ideas and to the notion that scientific progress and breakthroughs are achieved by overthrowing old ideas and old theories. It was hard for the scientists to visualize a totalitarian society allowing Copernican theory to be substituted for Ptolemaic theory, for example. Thus America's enemies represented the exact opposite of the kind of environment where the free exchange of information and ideas could occur.

Third, the racial policies of the Nazis gave added impetus to the mobilization of the scientists. In an American scientific community that now included large numbers of refugee scientists from Europe, a powerful additional motive was felt. Eastern and Southern European scientists, as well as both American and refugee Jews, saw themselves as particular objects of Nazi persecution.

In combination these factors provided powerful motives for basic scientists to move into applied work.[22] The scientists of World War II fame were generally people who, prior to the war, had been involved in developing scientific theory with no concern for its ultimate utility. That scientists were able to convert abstractions into weapons of great power would be one of the permanent legacies of World War II. Using their accumulated scientific knowledge, pure scientists became technologists of a unique and creative kind. The lesson would not be forgotten.

But more was involved. Not only did these scientists bring knowledge and theory to bear, they also infused the process with a uniquely open and experimental approach to the solution of problems. They were people whose professional careers had been oriented around developing new conceptualizations and experiments aimed at pushing the frontiers of knowledge. That mindset and experience made them uniquely adept at finding insights and experimental fixes that went beyond theory. They were people predisposed to thinking about doing things in new and different ways.

One humorous illustration of this mindset is recounted by Richard Bolt, a professor at MIT who was later a founder of the Bolt, Beranek, and Newman firm. Bolt was part of a group working on improving sonar. One of the key problems faced by that group was how to develop a more effective microphone that could work under the surface of the ocean. They saw great potential in the use of a salt crystal, but a central problem was how to keep the salt crystal dry. Bolt recounts that one day he was walking down the street in Cambridge, pondering how to protect the salt crystal from water, when he passed a drugstore. It occurred to him that one possibility might be to stretch a rubber condom over the crystal. He walked into the drugstore and bought a dozen condoms, which he took to the laboratory for tests. The tests indicated that the condoms offered real possibilities, so the next day Bolt walked back across the street and told the pharmacist he wanted another dozen. Bolt says he will never forget the look of shock on the face of that pharmacist.[23]

KEY ORGANIZATIONAL INNOVATIONS

The technological innovations of World War II would not have been possible without a uniquely creative community of basic scientists. For those scientists to be utilized, however, an act of organizational genius was required.[24] That genius was reflected in the way government mobilized this scientific talent and linked it to industry, building organizational networks. The organizational system had three ingredients: it frequently left the scientists physically located in universities; it was ad hoc rather than bureaucratic; and funding was open-ended and flexible.

The key decision involved the creation of a unique organization within the White House—the Office of Scientific Research and Development (OSRD), headed by Vannevar Bush, an electrical engineer and MIT vice-president. OSRD was modeled on an approach that had been recommended in a study of how to mobilize the scientific community to contribute to solving the problems of the Depression. In choosing the OSRD approach, President Roosevelt rejected the tack that had been used during World War I, when scientists had been inducted into the military and put under the direction of the Officer Corps. From the point of view of the scientific community, the World War I approach had two fundamental weaknesses: first, the rigid bureaucratic structure of the military services was seen as incompatible with the open, unstructured character of information exchange and contact required for creative science; second, the Officer Corps was generally scientifically illiterate and thus not sensitive to the needs of or not capable of understanding the substance of the scientific process.[25] For the formality, structure, and scientific illiteracy that characterized the effort to mobilize science during World War I, the OSRD approach substituted informality, lack of structure, and scientific literacy.

In Bush and the staff that made up OSRD—which included Arthur Compton and James Conant, both distinguished scientists—Roosevelt had some of the ablest and most widely known scientific/technical people in the country. Bush and his colleagues personally knew many of the most distinguished scientists in the United States. They were thus able to bring to their task a personal quality and a reservoir of personal trust that was of great value. Moreover, they understood the mores of the scientific community and the norms of its operation; thus they understood that the most fruitful way to utilize and mobilize the scientists was to give them a problem and get out of the way.

Bush and his colleagues acted primarily as recruiters, liaison people, and facilitators. Through their contacts with the military they identified technological problems and needs and then found the appropriate people to work on them. An example was the proximity fuse. Even before the United States entered World War II, the problem of proximity fuses had been under consideration by the United States Navy. The need was for a triggering device that would cause a shell to explode when it came within a given range of an aircraft or a certain distance above the ground. The problems of making this kind of fuse with a long shelf life, that could be produced in mass quantities, that was reliable, and that was also safe to handle were incredibly difficult. After a variety of options were eliminated, the major interest focused on a radio fuse. In his history of OSRD, James Phinney Baxter summarizes this effort as follows:

If one looks at the proximity fuse program as a whole, the magnitude and complexity of the effort ranked among the three or four most extraordinary scientific achievements of the war. Towards the end of hostilities it monopolized 25 percent of the facilities of our electronics industry and 75 percent of the nation's facilities for molding plastics. The job would never have been done without the highest degree of cooperation between American science, American industry, and the armed services. That it was done at all borders on the miraculous. The results are writ large in the history of war on land and sea and in the air.[26]

The people at OSRD not only picked up and facilitated the development of needs identified by the military but also identified and pushed the development of technologies initially rejected by the military. An example is the Dukw amphibious vehicle. Baxter tells the story of the Dukw as follows:

Hartley Rowe had discussed with Bush before the war the necessity of discharging cargo directly across a beach. His colleagues Putnam and Stephens came up with a solution. "If we could get an amphibian that would take cargo from a ship lying out in the harbor and bring it right to railroad sidings it would speed up shipping so much that it would be just as good as adding a million tons to the Allied Merchant Fleet." In amphibious operations moreover, which would soon be necessary on an unheard of scale, there would be no harbors or piers ready for use after allied bombings and gunfire and enemy demolition. A vehicle which could operate on both land and water over reefs and sandbars would prove invaluable in keeping cargo moving across the landing beaches.[27]

The vehicle, developed by General Motors, was built and demonstrated but, despite its good performance, the chief of the Corps of Engineers indicated there was no need for it because there was a widespread feeling within the Army that other vehicles could do everything the Dukw could do. Those in OSRD promoting it nonetheless managed to arrange another demonstration. Baxter describes what happened:

The Development of the Synthetic Capability

Four days before the demonstration a Coast Guard patrol boat with seven men went ashore at night on Peaked Hill Bars in a full gale and the Coast Guardsmen found conditions too severe to effect a rescue, either by surfboats or by a breeches buoy. Putnam, Stephens, and two Coast Guard officers headed a Dukw straight into the surf, cleared the breakers, reached the wrecked ship a quarter of a mile offshore. In six minutes they had rescued the seven Coast Guardsmen whose ship vanished during the night. When the Dukws were demonstrated four days later, 86 top-ranking officials hailed them with enthusiasm.[28]

By the end of the war there were seventy-six Dukw companies, each with fifty Dukws, in the U.S. Army and Marine Corps.

The OSRD approach was to go to research facilities and university scientists, define the nature of the need, indicate that it was urgent, and ask them to do whatever was needed to meet the requirements. Government would support the scientists by hiring personnel, providing them with the necessary equipment, and doing this without complex contracting, funding, or auditing burdens. Attention was focused on solving the problem. When those working on the project needed help, materials, equipment, or anything else, OSRD acted to support them. An illustration of the operating style is given by Arthur Compton of OSRD. Compton visited Edward Mallinckrodt, chairman of the board of Mallinckrodt Chemical Works of St. Louis. As Compton tells the story,

[W]e needed to process 60 tons of uranium. It was impossible to set a price until the process was worked out in more detail. The only assurance I could give Mallinckrodt was that the Office of Scientific Research and Development would supply him with a letter of intent to work out a contract that did not leave him financially the loser.[29]

Thus not only did OSRD mobilize university scientists, it also mobilized American industry and linked it to the university scientists and to the Navy and War Departments. OSRD was in the business of building the organizational networks necessary to carry out military hardware innovations. For the first time, university scientists, industrial people, and government officials became co-members of teams oriented to achieve objectives that could, at the time they were undertaken, be defined only in terms of performance needs. The proximity fuse resulted from the need for a triggering device. No one knew enough about how to make it to write detailed specifications. For the first time, therefore, the government was in the business of mobilizing the private sector to accomplish tasks that no one knew how to accomplish at the time they were undertaken.

Changes Wrought by World War II

The technological successes of World War II fundamentally changed expectations about what was necessary to maintain national security and about the roles and relationships of institutions. In the future, national security would require a capacity to carry out synthetic innovation of weapons systems continuously. Innovative leadership henceforth would require that government be continuously in the business of supporting research and development (R&D). In the first two decades following World War II, driven by the needs of the Cold War, R&D expenditures were to be the most rapidly growing portion of the federal budget. By 1965, for example, R&D expenditures would represent 15 percent of the federal administrative budget.[30]

A NEW ROLE FOR GOVERNMENT FUNDING

Something incredible happened during the first two decades following World War II. The expenditure of government funds for the purpose of understanding what was not understood and for developing products we did not know how to make became a huge enterprise. This pursuit of the unknown was budgetarily divided into three categories: (1) basic research—the pursuit of understanding without regard to social utility; (2) applied research—the pursuit of understanding because understanding was deemed necessary to achieve some social goal;

and (3) development—the design and construction of prototype products with superior or new performance capabilities. World War II had resulted in a social metamorphosis. It was now considered not only acceptable but in fact imperative that government financially underwrite research aimed at theoretical understanding as well as development of products with superior or new capabilities. Financial support of both research and development was necessary because theoretical understanding and new or superior technologies were now perceived to be inseparable and interdependent.

The great contributions of the World War II scientists had converted a view that had previously been restricted to the scientific community into the conventional wisdom of society—theoretical scientific discoveries ultimately have direct social utility. For years, scientists had told a story about Michael Faraday and a British prime minister. The story was that when the prime minister was introduced to Faraday, he asked Faraday what the use of his research on electricity was. Faraday was reported to have responded, "I know not, Mr. Prime Minister, but some day you will tax it." World War II made many of the leaders of government converts to a belief in Faraday's rule.

A NEW RELIANCE ON THE PRIVATE SECTOR

The second fundamental change to come out of World War II resulted from the fact that much of the productive science and technology had been carried out in nongovernmental research and development institutions: in universities and industry. As a result of the successes of these institutions, by 1965 roughly 80 percent of all federal R&D expenditures were going to nongovernmental institutions.[31]

The invention of the R&D contract during World War II, a contract that allowed government to buy research and development that it could not define in detail, represented a fundamental change. In combination, R&D contracts and procurement contracts were to provide the basis for linking universities, industry, a variety of other R&D institutions, and government into an organizational complex unlike anything that had previously existed. Now government had the capability to say, "We'll pay you to develop a system with the following performance capabilities. We don't know how those capabilities will be achieved. You achieve them. We'll pay you to do it. When you've produced prototypes that meet the prescribed performance standards, we will give you another contract for their production." In the years following World War II the federal government became the main patron of university science and engineering, providing 62 percent of all the support for university-based research and 83 percent of funding provided by non-university sources.[32] The graduate research university system became overwhelmingly dependent on government support; but it is equally true that government became dependent on the university system.

A similar pattern of interdependence developed between government and industry. Professor H. L. Nieburg describes the

> growth of government-industry contract relationships under which, in the words of David E. Bell [then director of the Bureau of the Budget], "Numbers of the nation's most important business corporations do the bulk of their work with the government." The Martin Company, for example, does 99 percent of its business with Government. Bell asked: "Well, is it a private agency or is it a public agency?" Organized as a private corporation and "philosophically . . . part of the private sector" yet "it obviously has a different relationship to government decisions and government budget . . . than was the case when General Motors or U.S. Steel sold perhaps two or five percent of their annual output to the government."
>
> . . . The Contract state of the post-war world must be viewed as a drastic innovation full of unfamiliar portents.[33]

World War II fundamentally changed the attitudes of government leaders, members of the scientific community, and industry leaders; and it saw the invention of government contracting arrangements that were to tie these three sets of institutions into a dynamic new complex, a complex capable of carrying out continuous synthetic innovation.

The experience of World War II especially changed the perceptions and expectations of the leaders of the military services. All thinking about future wars had to assume that military capabilities could be fundamentally changed by new synthetic innovations. Thus, so far as the military was concerned, the major consideration was how to maintain and enhance the capability for innovation that had been developed during World War II. The Cold War significantly reinforced the government's need to maintain a tight linkage with the scientific community and American industry. World War II triggered a similar change of attitudes in industry. During World War II, government had been the economy's largest buyer and consumer, a pattern perpetuated with the beginning of the Cold War. Now the route to the government market was leadership in innovation. Companies involved in developing prototype military technologies were also the ones that became the suppliers of the military. Then another important factor began to emerge. As a pattern of spinoff commercial products became increasingly apparent, the value of being part of the national defense organizational complex that was carrying out synthetic innovation was enhanced. Not only did military innovation provide access to the large government market, but it also increasingly provided competitive advantages in the commercial market.

For the university research community the lessons of World War II were twofold. First, university researchers were impressed with the great strides that could be made with government funding. Second, the development of new contracting arrangements during World War II demonstrated that accepting government funds need not result in the research community's losing control over the direction of research. Using the research and development contract and grant instruments invented during the war, university scientists were able both to define what work should be done and to determine how funds to support that research should be allocated.

In sum, World War II created a consensus view that there were great advantages for universities, government, and industry in maintaining close, intimate cooperation. The pattern of interaction that evolved during the war among these three sectors would have been unthinkable before the war. Then, following the war, the patterns of cooperation were both refined and expanded. Expansion was possible for two reasons: the close cooperation among organizations had proven its worth in meeting the Axis threat; and the Cold War provided a new threat. By the end of World War II the defense complex had inertia on its side. Thus the only remaining need was to construct a rationale that made the new organizational arrangements fit with the rhetoric of the secular trinity.

The arrangements that made it possible for government to pay for R&D that was carried out in the private sector and then to buy its products were interpreted as a vital part of the free-enterprise system. The research and development contract provided the legal mechanism necessary to maintain that pattern of fruitful cooperation; it is discussed in the next chapter. While the secular trinity continued to dominate the nation's rhetoric, a very different reality was being evolved.

CHAPTER 7

The Contract
Federal System

WHAT had been launched as a set of ad hoc organizational responses to immediate needs during World War II was converted into a large, regularized organizational complex in the postwar period. World War II innovation was a response to a crisis characterized by the special creativity and energy that crises often call forth. On the other hand, the synthetic society does not depend on such urgent and special motivators. Rather, continuous synthetic innovation in the post-war period has relied on much more mundane and predictable factors. Specifically the synthetic society has used organizational incentives and forms invented in the defense sector and exercised by government.

World War II was to impact on future innovation in many ways, but, as Don Price, the former dean of Harvard's John F. Kennedy School of Government, has stated,

> [T]he most significant discovery or development . . . was not the technical secrets that were involved in radar or the atomic bomb; it was the administrative system and set of operating policies that produced such technological feats.[1]

For the last four decades the defense sector has been the nation's largest source of synthetic innovation, and it remains the overwhelming focus of government R&D as well as the employer (directly or indirectly) of between 17 and 30 percent of the nation's scientists and engineers.[2] Defense innovation has been a success story. Throughout the Cold War the United States has continually enjoyed the benefits of weapons systems with performance capabilities superior to those of the Soviets. Consistently the United States has been the force driving the innovation-to-obsolescence cycles of weapons systems, and consistently the Soviets have played catch-up.

How was it possible to build, and how has it been possible to sustain, the organizational complex that has produced this record of innovation? Does the defense innovation experience provide a model for successful commercial innovation?

The story of defense innovation is inextricably intertwined with what Don Price calls "the contract federal system"—i.e., the system held together by contracts.[3] It is a story about a complex of government and private-sector organizations that have become so interdependent that they cannot exist without each other. Indeed, the interdependence has become so pervasive it is no longer possible to distinguish public from private-sector functions.

Defense innovation has required tapping the full range of scientific/ technical expertise. That expertise is distributed among six types of organizations that make up the contract federal system: federal agencies, federal laboratories, federally funded research and development centers (FFRDCs), nonprofits, universities, and industry (see figure 7–1). Continuing innovation requires close and intimate cooperation among these organizations. The contract federal system has created an environment in which the diverse organizations that make up the defense complex share a common need: they need to innovate to prosper. It is the genius of the defense complex that the well-being of the individual organizations has been made to coincide with the nation's need for continuous weapons systems innovation.

The drive toward continuous innovation in defense results from its having been a high national priority for decades. From the beginning of the Cold War, U.S. national security has rested on an ability to compensate for the larger numbers of Soviet military personnel and hardware with superior weapons systems. Performance has been critical, whereas cost has been a secondary concern. Only an organizational

FIGURE 7–1
The Contract Federal System

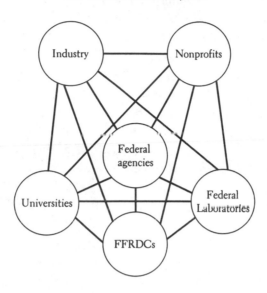

system that could design weapons with capabilities that were beyond the state of the art and deliver those weapons ahead of the Soviets could provide the nation with a credible defense. The organizational complex was built with the implicit assumption that the United States and the Soviet Union would be involved in innovation-to-obsolescence-cycle competitions in perpetuity. To lose in any competitive cycle was unacceptable. With the nation's survival at stake, there was no need to conform to the dictates of the secular trinity in anything other than a rhetorical way.

The capability for continuous innovation required that the contract federal system provide three sets of incentives. First, incentives had to be created that made each organization in the complex strive continuously to push the state of the art in its area(s) of expertise or specialization. Second, it was necessary to create incentives that would cause diverse organizations with very different motives (government, industry, and universities) to interact and cooperate with each other in the pursuit of objectives none of them could accomplish alone. Third, it was necessary to create incentives that translated that integration and cooperation into a synergistic capability. The complex of organizations, on a continuing basis, had to deliver innovations that represented more than the sum of the individual organization's capabilities. The creativity required for innovation had to be made regular and routine.

Complex innovation—and modern weapons systems are surely among the most complex—are developed by integrating the successes of a large and diverse process of trial and error. Of course, progress by trial and error inevitably leaves a large trail of technical failures. The trick in achieving continuing innovation is to get groups to think creatively and organizations to take the failure of a large number of those creative thoughts as routine, expected, and even desirable. Continuous innovation in defense has required standing on its head the conventional wisdom about the behavior of groups and large organizations. There have been two key ingredients to success. First, the complex decoupled financial cost from technical failure. Second, it has provided high levels of confidence that major rewards will come from success, with success defined in technical performance terms. The instruments used in mobilizing, organizing, and driving the organizations that make up the complex are government contracts of two specific types: the research and development (R&D) contract and the production contract. The R&D contract is the key to continuous innovation, and the production contract is the "pot of gold" at the end of the innovation process.

R&D contracts are used to pay for the discovery of new knowledge and the development of new capabilities. In concept, all defense research and development expenditures have as their ultimate goal the production of prototype weapons systems—a new tank, a new cannon, a new airplane, a new submarine, a new missile, and so on. Prototype weapons systems are developed with the explicit goal of providing capabilities that are superior to or beyond anything that existed before. Once the prototype has been tested—ideally under field conditions—and has met the performance criteria, it is traditionally a near-certainty that the organizations involved in developing the prototype will be the recipients of the production contract.

The organizations that develop the prototype are assured of the production contract because, in the process of development, that network of organizations will have accumulated experiential learning. Only that unique body of experiential learning acquired in developing the prototype allows for the rapid movement to production. Any effort to move production to a new organizational network would be costly because it would require repetition of at least part of the learning curve experienced in the development. Thus it is substantive, experiential knowledge acquired in carrying out research and development that guarantees the follow-on production contract.

The R&D Contract: Basis and Incentives

Understanding the unique set of incentives that are central to the contract federal system requires a look at defense R&D expenditures. Since national security rests on superior technology that results from research and development, large R&D expenditures are a permanent fixture in the defense budget.

Although R&D monies are a given, few of the organizations that make up the defense complex can take receipt of those monies as a given. This is because, unlike most government expenditures, R&D funds largely are not tied to particular organizations on some line item or long-term formula basis. Rather, the overwhelming portion of the roughly $40 billion spent on defense R&D in 1988 was distributed in the form of tens of thousands of competitive grants and contracts.[4] Each grant or contract involves an agreement between the funding agency and the R&D organization to carry out some defined work. Some of the contracts involve a commitment by the agency to spend a fixed amount of money and a commitment by the contractor to do as much as it can with that money. In this case, the ultimate output is uncertain. Some contracts involve a commitment by the agency to fund the contractor until a particular task is achieved or capability developed. In this latter case, the ultimate amount of money and time expended is uncertain.

Whatever the precise form, the R&D contract process has two unique characteristics. First, the primary criterion for choosing one potential contractor over another is scientific or technical competence, although scientific/technical merit as the basis for allocating R&D

monies needs to be examined carefully. It does not mean that the selection process is free of traditional political pressures. Rather, it means the "pork barrel" process controls selection only when the potential contractors are all judged to have the necessary scientific/technical qualifications. Second, R&D contracts are always in the process of coming to an end. To put it another way, recipients of R&D contracts are always in the business of working themselves out of a job. Those contracts that involve fixed funding for a specific time period are clear-cut. Those that call for the development of particular performance capabilities may leave the cost and time unclear, but they are also dominated by the same central fact—once the performance criteria called for in the R&D contract are met, the contract terminates.

R&D contractors, whether they be professors in universities or large industrial corporations, are therefore perpetually involved in a cycle of completing work under contract and writing proposals for or competing for future work. The key to successful competition is an assessment by the funding agency that the bidder or proposal writer has the competence to carry out the next research-and-development task. Thus if one looks at the defense contract federal system, one sees that every year tens of thousands of R&D contracts are terminated and tens of thousands of R&D contracts are initiated. The same organizations are involved in writing proposals and bidding, carrying out research and development, and terminating research and development activities.

As a result, the contract federal system creates a set of incentives that are central to continuous innovation. First, the ability to compete successfully for future contracts rests on maintaining state-of-the-art scientific/technical competence in an environment where the state of the art is always changing. Given the perception that the organizations that have reputations for being the most advanced and creative have a competitive advantage, a focus on the future becomes essential for survival. Organizations that are a part of the defense innovation complex are competitors in a de facto "futures market." The best-known futures market deals with agricultural commodities. Winners in that agricultural market are those who make the best estimates of supply/demand relationships months into the future. Of course, success or failure may be determined by variables such as weather, government policy, pests, wars, and so on. In the defense innovation futures market, the future is years away and winning rests on having the appropriate scientific/technical capabilities needed to contribute to the next incremental or radical innovation.

Competition in this futures market is complex. Those who will compete for the next R&D contract are also, because of the system's seamless character, inevitably heavily involved in defining the performance criteria that will be called for in the contract specifications. The defense complex is a system that processes ever-evolving information about technical capabilities; consequently, the ability of an organization to push the state of the art not only contributes to its reputation and thus its competitiveness when contracts are allocated, but its advanced work is likely to be written into the performance criteria established by the contracting agency. Clearly the greatest competitive advantage is to have specified performance criteria that only your organization can meet.

The failure to win a contract in any innovation-to-obsolescence cycle is viewed not just as a loss in that cycle but as a liability that grows exponentially. Since technical capabilities are heavily experiential and organizational learning is the cumulative result of trial and error, not being a participant in every cycle involves an exceptionally high risk. Essentially the organization becomes less and less competitive as it fails to participate in each cycle, and the state of the art moves away from the organization's capabilities.

Another incentive is engaged once an R&D contract is received: having successfully competed for an R&D contract, organizations operate within a uniquely secure environment. R&D contracts provide incentives for group creativity and organizational trial and error because they decouple financial costs from technical failure. Outside the contract federal system, technical failures carry financial costs, but making technical failures low- or no-cost activities creates powerful incentives for experimentation. In truth, given the need to push the state of the art in order to be able to win in the future, organizations have strong incentives to carry out broad-based experimentation. Experimentation is both the route to successful completion of existing contracts and the substantive basis for the organizational learning that provides the ability to compete in the future.

Evolution of the R&D Contract

The development of the organizational complex that has carried out weapons innovation since World War II represents a fundamental historical transition in the evolution of the American political/economic system. The R&D contract was a key instrument used to develop this new system, so to appreciate what occurred, an historical look at this contract is necessary.

The R&D contract was invented to meet the needs of World War II. These needs were described in chapter 6 from the standpoint of the inadequacy of traditional interrelationships between the federal government and R&D performers in the private sector. Likewise, as the prospect of U.S. entry into the war became ever more real, traditional federal contracting practices were inadequate to the nation's needs.

Prewar contracting practices were designed to get the lowest price and to protect against dishonesty on the part of federal officials and collusion on the part of private-sector contractors. Prewar contracting required that the government provide detailed specifications for whatever it wished to buy. In a Brookings Institution study, Clarence Danhof has summarized the contracting process.

> Government agencies could normally procure goods and non-personnel services only by a) public advertising for bids responsive to detailed specifications; b) public opening of bids at a specified time and place; and c) award of the contract to the lowest responsible bidder complying with the conditions of the advertisement for the bids.[5]

Congress recognized the obsolescence of these procedures and in 1940 passed the National Defense Expediting Act, under which the military "services were granted the authority to buy through negotiated contracts involving either a fixed price or a cost-plus-fixed fee."[6] This authority was to be the basis for the great flexibility of the R&D contract. Initially both the War and Navy Departments were slow to use the new contractual freedoms. By the end of the war, however, the basic form of the R&D contract had been evolved and the defense agencies had gained experience in using it.

The R&D contract provides a legal basis for agencies to buy products and services that they cannot define in detail. The contract establishes a relationship between a federal agency and a private-sector organization whose goal is to develop a capability for accomplishing tasks that are beyond the state of the art. As was implied in chapter 6, it is an agreement whereby both parties work together to accomplish performance goals no one knows how to accomplish.[7] In fact, even the performance goals in R&D contracts are subject to continuous modification because evolving scientific/technical capabilities change the prevailing views of what it is possible to accomplish. Initially an R&D contract may call for delivering certain capabilities within a fixed period of years. By the end of that period, technologies may have been developed that offer greatly enlarged capabilities. Alternatively, accomplishing the performance called for in the contract may not be possible because of unanticipated scientific or technical problems. Thus, although R&D contracts have as their central purpose the achievement of performance objectives that are beyond the state of the art, those objectives are commonly modified in the course of carrying out the innovation.

Harold Orlans, another Brookings Institution researcher, has characterized the R&D contract and compared it with the traditional contracting relationship as follows:

In a "hands-off" buyer-seller relationship, all of the terms that are necessary to the satisfactory completion of a contractual obligation may, perhaps, be put down on paper: the exact goods to be delivered, at a given time to a given place for a stated sum. But, the most significant aspects of research, development, and managerial contracts are precisely those that cannot be satisfactorily reduced to writing for they involve the quality of effort made to reach, not a precise target, but goals that must be continually readjusted by mutual agreement; and—beyond gross and obvious limits—the balance of initiative and responsibility, of freedom and control, that should lie in public and private hands is also a shifting one that must be worked out individually for each major contract.[8]

In sum, the R&D contract involves a set of arrangements that allow both the funder (the federal agency) and the doer (the private-sector organization) to agree they will work together intimately to do something neither of them knows how to do under circumstances in which their various functions are subject to change. Making a separation between the public sector and the private sector is not possible in these contractual relationships. Thus the separation that is central to the concept of a free market as prescribed by the secular trinity makes no sense. On the contrary, intimate and informal public- and private-sector interactions work best in the R&D contractual relationship. Both the contractor and the contracting agency must understand in an intimate way the problems, needs, and circumstances faced by each other.

In its earliest and simplest form the R&D contract defined the objectives to be achieved and specified three sets of financial terms. First, it committed the agency to pay the direct costs of carrying out the R&D defined in the contract. Second, it involved an agency commitment to pay overhead costs (e.g., buildings, utilities, administrative expenses), usually some percentage of the direct costs. Third, it involved a commitment to pay a fee or profit. Normally the R&D contract had a fixed fee, and profit was constant regardless of costs. Contracts with these terms are known as "cost plus" contracts.

In the early post–World War II period, R&D contracts were frequently both "cost plus" and "sole source." In the case of sole source contracts, the agency issuing the contract selected the organization believed to be most capable of carrying out the task and negotiated the terms of a cost-plus contract. Other potential contractors had no opportunity to compete in the sole source context. These arrangements allowed agencies to move rapidly to buy both the pursuit of understanding and the development of capabilities from the organization or organizations judged to be most capable. A typical large development contract would specify the performance objectives for a technology. In the case of a new fighter airplane, for example, the contract might specify its speed, the distance it should be able to fly without refueling, the ordnance it would carry, the characteristics of its avionics, and so forth, and the contractor would commit to develop that system.

In the immediate post–World War II period, defense agencies had great contractual freedom and could act with speed in negotiating such contracts. Over the course of the postwar period, however, contractual flexibility has been constrained. As problems have been identified in the contract federal system, the response has been to seek greater efficiency through changes in contract terms, use of competitive bidding, tighter management, and more audits. Thus the answer to R&D contracting problems has consistently been sought in the secular trinity's insistence on efficiency. In chapter 9 the effort to apply ever more detailed and rigid contract terms and rules and the threat they pose to innovation, are discussed. At base the genius of the defense complex has been its ability to respond in a rapid and organizationally ad hoc fashion to substantive needs. The huge size, great complexity, and dispersed character of the expertise required for defense innovation precludes tight, centralized management.

Three agencies sit at the center of the defense contract federal system: the Department of Defense (DOD), the National Aeronautics and Space Administration (NASA), and the Department of Energy (DOE). Together these agencies control and allocate the overwhelming majority of all federal R&D monies. Those monies, allocated through contracts, are the primary levers the agencies use to conceptualize, design, and develop the nation's ever-changing menu of weapons systems. These agencies do not act as a monolith, nor are they guided by a unified set of priorities. Defense technologies are not derivatives of some broad strategic or tactical doctrine. Neither is doctrine determined solely on the basis of evolving technological capabilities. Rather, the weapons systems are an amalgam of evolving and interdependent technologies and doctrine. This is inevitable given the size, complexity, and number of innovations funded by the defense agencies. In addition to the roughly $40 billion R&D budget in 1988, the Department of Defense spent approximately $80 billion on weapons production.[9]

In reality, the defense organizational complex is a holding company for what are normally separate weapons-development organizational networks. For example, within DOD, responsibility for R&D is divided among five components: the Army, the Navy, the Air Force, the Defense Advanced Research Projects Agency (DARPA), and the new Strategic Defense Initiative Office (SDIO). Furthermore, within each of these organizations—but particularly within the three services—

responsibility for R&D is divided among diverse commands, offices, and programs that have varying degrees of authority and discretion at any point in time. Specific organizational units exist or are created for carrying out everything from basic research to the development of specific weapons systems—e.g., missiles, tanks, airplanes, and so on.

Although created as a civilian agency, NASA has, since the end of the Apollo program, been involved increasingly with and linked to military programs. What direction NASA will take in the post-Challenger era remains unclear. With a growing budget now totaling roughly $6 billion, NASA is a major factor in the contract federal system.[10]

About half of the 1988 budget and a large portion of the personnel and research facilities of the DOE were devoted to that agency's military/nuclear activities.[11] The department has responsibility for the design, development, and production of the nuclear weapons that are integral to the mission of the DOD.

Coordinating the interdependent activities of the three agencies is a complex task, one that can only be accomplished by teams and committees of experts arriving at consensus recommendations that go to all three agencies. Defense innovation evolves; it is not directed. This point is illustrated by the SDI. Utilizing research-and-development funding from all three agencies, government laboratories and private-sector R&D organizations had, by the late 1970s, made significant advances in such technologies as directed- and kinetic-energy weapons, multi-spectral sensors, large-scale power sources, space-launch capabilities, and high-performance computers and communications. These and other advances led to discussions among members of the defense science/technology community about the possibility of a new, vastly more capable ballistic missile defense supersystem. One of the participants in that discussion was Edward Teller, a distinguished scientist who has broad and continuing contacts throughout the defense innovation organizational complex. It is reported that Teller communicated the possibility of a strategic defense capability directly to President Reagan and was influential in convincing the president that the nation should adopt such an initiative.[12]

Neither the president nor the technically sophisticated Teller had a detailed understanding of how a workable strategic defense capability might be built. No individual can understand, except in a general way, a system so complex that it requires pushing far beyond present capabilities. Teller's role, then, was to communicate a technological possibility or vision; this vision was itself the conceptual creation of an organizational complex that exists to push the state of the art of defense technology continuously. The decision to pursue the SDI led to the creation of a new network. Some sense of the challenge posed by the SDI is suggested in the following summary by C. A. Zraket, president of the MITRE Corporation:

> Building a strategic defense against nuclear ballistic missiles involves complex and uncertain functional, spatial, and temporal relations. Such a defense system would evolve and grow over decades. It is too complex, dynamic, and interactive to be fully understood initially by design, analysis, and experiments. Uncertainties exist in the formulation of requirements and in the research and design of a defense architecture that can be implemented incrementally and be fully tested to operate reliably.[13]

The decision to undertake the Strategic Defense Initiative has the same two components that are associated with most synthetic innovation. First, it involves a decision to do what has never been done before and what no one knows how to do. Second, the decision is the result of a seamless web of interaction and communication among federal agencies and those organizations that will ultimately participate in developing, producing, and deploying the innovation. Defining the performance standards for the next innovation cannot be done independent of those who will develop the innovation. In the defense contract federal system it is not possible to distinguish between the public and the private sectors. In a formal sense government may be the definer of what is needed. In reality, recipients of federal contracts will have helped define the content of the contracts.

CHAPTER 8

The Organizational Complex

FOR SEVERAL REASONS, understanding how the defense organizational complex developed and has evolved is important to developing the capacity for commercial innovation. First, a large number of today's high-technology commercial products have some of their roots in the defense complex. This complex developed the basic organizational and procedural arrangements needed to accomplish synthetic innovation. Thus the synthetic society has its roots in the defense innovation system. The defense complex, then, offers insight into what is necessary to develop an organizational complex that is capable of innovating for the export market. Second, the defense complex is a major consumer of the nation's scientific/technical resources, and any effort to refocus the nation's innovative capabilities will require moving some of those resources to other sectors. Finally, the defense complex is still the source of technological spinoff to the commercial sector.

Perhaps no message about the synthetic society deserves more emphasis than the need to structure organizations and organizational networks around substantive needs or goals that require pushing the state of the art. Defense innovation requires the ability to tap and gain synergism from the necessary expertise and skills wherever they may be located. When innovative ventures are undertaken, there may be a less-than-clear picture of the precise sets of expertise that will be needed. Since innovation is, in part, empirical and trial-and-error in character, organizational networks must be capable of linking to and decoupling from organizations as they demonstrate an ability to contribute or, alternatively, as they demonstrate that they cannot contribute.

Continuous synthetic innovation in defense, then, requires organizational flexibility at both the level of the network and the complex. R&D contracts are used to ensure that the organizations that make up the complex are continuously pushing the state of the art, and they are also used to build and modify the networks required to carry out specific innovations.

In what follows we examine the texture, diversity, and flexibility that have characterized the defense contract federal system by looking at five generic types of R&D contracts: (1) the research grant, (2) the master contract, (3) the special-purpose contract, (4) the facility management contract, and (5) the prime contract.[1]

Research Grant

Research grants serve primarily to ensure the continued advancement of understanding in an area and are not normally linked to a specific innovation goal. The grant is generally used to provide funding for individuals or small research groups at universities. Although formally a contract between an organization (e.g., a university) and a federal agency, the grant is a de facto contract between a specific researcher or group of researchers and a federal agency. The agency provides a fixed amount of money to cover all or some of the researcher's costs, and the performing organization agrees that only the specified researcher will use the funds. The researcher agrees to work in a designated subject area and to publish the results. The grant is the contractual manifestation of a belief in "Faraday's rule," discussed in chapter 6. That is, it allows government to underwrite basic research, the direction of which is left in the hands of the research community, on the assumption that the research will have ultimate societal payoff even though what that may be is presently unknown.

The grant selection process normally works as follows: the agency announces a desire to support research in broad areas such as physics, chemistry, and so forth and invites researchers to submit proposals. Thus the definition of the research is left in the hands of the researchers. Once investigator-initiated proposals are received, the agency selects a panel of experts from the research area to review the proposals. Normally members of these "peer review" panels are researchers from nongovernmental organizations—mostly universities. The panel assesses the proposals' scientific merit. In some instances review panels rank-order proposals and the agency funds on that basis. The grant, then, allows the research community to both define research directions and select those who will do the research. Using the grant, government has become the patron of university scientists, a patron who gives them great freedom and leaves the judgment of their performance to their research peers.

Although the grant is only one of several contract forms used to support university research, it is widely believed to be the most important to the maintenance of U.S. leadership in the basic sciences and in fundamental engineering research. Federal funding has tied large numbers of university researchers into the organizational complex that carries out defense synthetic innovation. Universities received on average between 63 and 69 percent of their R&D funding from federal agencies during the 1970–1986 period.[2] The three defense-related agencies (DOD, DOE, and NASA) provided 50 percent of non-health-related federal R&D funding going to universities in 1985.[3] This funding, plus a nearly equal amount from NSF, is believed to provide the defense innovation system with three things. First, it generates the new discoveries, ideas, and information that fuel weapon systems innovations. Second, the fundamental research carried out in universities provides the training ground for those who carry out R&D in the other organizations that make up the defense complex. Third, university researchers serve as expert advisors and consultants to all the organizations that make up the defense organizational complex.

Although the grant was created to insulate basic research from a need to demonstrate its social utility, it is an act of faith in long-term utility. The rapid evolution of innovation-to-obsolescence cycles in defense, and the resultant perception of a compression of time between basic research and utilization, has made the link between basic research and application an almost universally accepted notion.

Finally, while grant-supported research is guided by the research community at the specific project level, federal agencies do choose the emphasis given to various areas of science and engineering. For example, the Reagan administration generally gave more emphasis to the physical sciences and engineering and less to the life sciences.[4] One explanation is that the physical sciences are seen as more relevant to defense innovation.

The grant gives basic researchers great freedom, with their only obligation being to publish their results. In fact, no such obligation is necessary since the research community rewards its members through recognition, and that recognition is itself based on publication.[5] In sum, the grant supports the research community in doing what it wants to do (discover new knowledge), which is what the defense agencies also want—for the next generation of weaponry may be hidden in some innocuous basic research.

Master Contract

The master contract is used to create and sustain organizations devoted to R&D and/or technical management. Under this contract an agency provides sustaining support for a nonfederal organization, which in turn agrees to push the state of the art in some scientific or technical area or to provide oversight or technical management capabilities for innovations. Master contracts are frequently used when there is a need for integrating a variety of specialties or disciplines so as to provide a new area of competence.

Master contracts may be signed with any nongovernmental organization, but most have involved universities and nonprofits. Master contracts normally cover direct R&D or technical costs plus indirect or overhead costs. Contracts with nonprofits, as opposed to universities, also include fees. Fees give nonprofits the financial flexibility necessary to maintain a state-of-the-art scientific/technical competence. Fees also provide the discretionary funding needed to allow employees to carry out research and technical activities beyond those covered in the contract. An implicit assumption is that the nonprofit may identify through its own research initiatives findings that later will be of direct value to the agency. Even in the absence of identifiable serendipitous benefits, fees are assumed to produce organizations with increased substantive competence that benefits the funding agency.

Master contracts with universities usually result from a perception that there is or will be a future need for work in generic areas that are interdisciplinary in character. The goal of the agency is to tap the specialized disciplinary competence of universities and focus that competence on an area of interdisciplinary concern perceived to have potential utility.

Two examples of master contracts with universities suggest the character of work that is normally sought. One example is the program, initiated by the Department of Defense and then transferred to the National Science Foundation, which provides continuing support for Materials Research Laboratories. Here the perception was that great benefits could be gained from pushing the state of the art in materials. While the agencies may not know how they will utilize the work coming out of the Materials Research Laboratories, there are so many uses for new materials that it is assumed the work will have utility ultimately. The second example is the Engineering Research Centers (ERCs) recently established by the National Science Foundation. The ERC program is creating within universities cross-disciplinary capabilities that focus on areas such as telecommunications and biotechnology. The goal is that the ERCs will make a contribution to the nation's economic competitiveness in the future. A similar effort by DOD is the University Research Initiatives Program. The normal pattern is to provide multiyear funding for the centers, which are subjected to a performance review toward the end of the funding period. If the centers are judged to be making satisfactory progress toward their goals, funding is usually continued for another set period of years.

Master contracts also have been used both to create and to support a range of nonprofit organizations. As with the universities, the purpose normally has been to build and sustain a unique interdisciplinary capability. One type of contractual relationship is illustrated by the Aerospace

Corporation, which was created to provide systems management/technical direction capability. This corporation is actually the third phase of an organizational evolution that began in the 1950s, when a newly created but technologically weak Air Force was faced with the most formidable of technological challenges, the rapid development and deployment of the Atlas intercontinental ballistic missile system. Convair won the Atlas contract. Lacking an in-house capability to provide technical oversight and management for the Atlas program, the Air Force used a master contract to create a new profit-making company.

Two engineers previously with the Douglas Aircraft Company, Simon Ramo and Dean Woolridge, created the Ramo-Woolridge Corporation to serve as the Air Force's instrument for systems management and technical direction. A highly qualified scientific/technical team that integrated the range of skills and expertise required to provide oversight for Atlas was quickly assembled. Ramo-Woolridge was extremely successful. Its range of activities soon expanded far beyond the Atlas program.[6] Then, over time, Ramo-Woolridge developed a substantial capability for innovating components and subsystems, which made Ramo-Woolridge a competitor for providing components and subsystems for other major weapons systems. Other contractors argued that Ramo-Woolridge had an unfair advantage.[7]

To deal with the conflict between its systems management and production activities, Ramo-Woolridge spun off the systems-management, technical-direction function into a separate subsidiary called Space Technologies Laboratory. This proved an unsatisfactory response to the conflict-of-interest concerns, and the Air Force chose to have Space Technologies Laboratories converted into Aerospace, a separate, nonprofit organization incorporated under the laws of California. With this entity now a nonprofit, the need to return profits to investors was eliminated, so that the incentives for going into the hardware production business were also eliminated. Aerospace and similar systems-management, technical-direction nonprofits have played a continuing role in the defense complex by linking basic research to production and deployment capabilities. They are free of the profit motives that drive traditional corporations and are better able than universities are to organize interdisciplinary teams and deliver products and services on time.

143

A second type of nonprofit supported with master contracts is the "think tank." The best known is the RAND Corporation, originally created at the end of World War II by the Army Air Corps.[8] In its original incarnation, RAND existed as a separate project within the Douglas Aircraft Company. The project grew out of the World War II experience and reflected a desire in the postwar period to maintain a linkage between the civilian scientific/technical sector and the military. The goal of the Army Air Corps was to gain continuing access to the kind of scientific/technical talent that had been available during World War II, particularly that which had come out of the universities. Although at its inception RAND's purpose was only very generally defined, within a few years it had evolved to a point where its primary function was to study the combination of weapons systems and strategy necessary to win future wars. During World War II, strategic and tactical thinking had become the focus of civilians who sought to maximize the effective use of technological innovations using operations research methodologies. The RAND Corporation carried this genre of study into the postwar period and evolved a new methodology called systems analysis, which allowed RAND to go beyond the best utilization of available technologies to the definition of technologies that would make the maximum contribution to national security.[9] Given an organizational complex capable of synthesizing new technologies, the think tanks could define desired weapons systems that the complex could then develop. With the appearance of the think tank the defense innovation system had taken on dazzling new characteristics. RAND's contribution to the Air Force's strategy and tactics quickly led the other services to use master contracts to create think tanks. With the creation of the think tank, defense strategy and hardware development were linked into a seamless system. Doctrine and hardware became interdependent.

Credible study of future hardware and doctrine needs has required that think tanks have exceptional independence even while they remained the financial creatures of agencies. The degree of independence established by RAND and other think tanks at particular points in time is quite striking. That independence comes from the agencies' heavy dependence on the think tanks. Recall that the process of synthetic innovation requires building organizations that have unique experiential knowledge. Think tanks have become critical repositories of such organizational learning. In other words, the independence of think tanks is a direct result of organizational interdependence. Their independence has allowed the think tanks to play the role of "house heretics" or critics. Think tanks can serve as well-informed, state-of-the-art critics of agency weapons systems and doctrine, with the basis of the criticism being their conceptualization of superior systems and doctrine. Think-tank criticism frequently generates strong objections from those within the agencies who have a deep commitment to existing technologies and doctrine. Nonetheless, over time the agencies have continued to support these "house heretics" because they have provided a critical corrective agent within the defense complex. The great benefit of the think tanks is inherent in the word *house* heretic. The heretical criticism goes to the agencies directly. Thus the agencies benefit from knowledgeable critiques of their programs without those critiques being available to outside critics. In a context of ever-impinging obsolescence, continuous (but controlled) criticism is the elixir of survival.

The master contract has made it possible to build and support scientific/technical organizations with new capabilities. Once those organizations have acquired expertise, the agencies become as dependent upon them as the organizations are on the agencies.

Special-Purpose Contract

The process of continuous synthetic innovation in defense regularly identifies problems that require special studies or the development of specific technologies by organizations with established expertise. Within the defense community, specific individuals, groups, and organizations have a reputation for skill and creativity in particular areas. Given a technical community consensus on those people or organizations most likely to provide a solution to a given problem, agencies traditionally use sole-source, cost-plus contracts to meet their needs.

The ability to contract with rifle-like accuracy for the particular skill or capability needed to solve a problem is a critical component of the organizational complex that carries out defense innovation. Since synthetic innovation occurs by trial and error and involves many failures or partial successes, flexibility is critical to the innovation process. The complex must be able to tap needed expertise rapidly and integrate it into the process of synthetic innovation. The other side of this flexibility is the capacity to terminate lines of investigation and development efforts that prove unproductive.

Contracting in which the basis for selecting the contractor involves a judgment of his efficacy in doing what has not been done depends upon using the expert community's accumulated wisdom. The special-purpose contract, then, differs from the master contract in that the need is not to build new private-sector organizational capability but rather to tap existing expertise rapidly in relatively contained or identifiable areas.

Facility-Management Contract

Beginning in World War II, agencies identified the need for previously nonexistent organizations with capabilities for developing or improving complex technologies. As we have seen, scientific/technological complexity increases as a function of the number of different areas of expertise and skills needed to carry out the task. The most complex World War II innovation was the atomic bomb, produced by the Manhattan Project. While carrying out that project, the Army Corps of Engineers made extensive use of facility-management contracts to create new organizational capabilities. The Corps negotiated sole-source, cost-plus contracts with private-sector organizations to build and manage new facilities.

The use of facility-management contracts has been a distinct feature of the postwar defense-innovation system. Under facility-management contracts the agency pays all the capital costs (building the facility from the ground up) and all the operating costs, and pays the contractor a fee for building and then managing the facility, while the agency retains ownership of the physical facilities. Today these facilities are often referred to as federally funded research and development centers (FFRDCs). In the post–World War II period, the Atomic Energy Commission (AEC)—the agency that replaced the Army Corps of Engineers in jurisdiction over atomic energy—used facility-management contracts to manage and expand the bomb laboratories into large, multipurpose national laboratories and to create new facilities.[10] Oak Ridge, Argonne, and Los Alamos National Laboratories represent major facilities operated under facility-management contracts held by Martin Marietta, the University of Chicago, and the University of California, respectively.

Although facility-management contracts have parallels with master contracts, they generally differ in function. Facility-management contracts are normally used to create permanent organizations with the capability to develop and/or produce and/or operate complex hardware systems.[11] Facility-management contracts then offer the dual benefits of government ownership of facilities along with the operational flexibility of private-sector organizations. Such arrangements have been preferred over building in-house R&D facilities for a variety of reasons. First, federal regulations, governing agency personnel and operations alike, are generally seen as barriers to flexibility. Second, civil service pay scales frequently make it difficult to recruit the best scientific/technical personnel, particularly in high-demand specialties; even where that is not the case there are always mandated ceilings for the number of civil service employees in an agency.

The range of federally funded research and development centers can be illustrated by looking at two cases. One is the Sandia National Laboratory. Sandia came into existence in 1949 when the AEC determined that a new bomb development facility needed to be set up near Los Alamos Laboratory. David Lilienthal, chairman of the Atomic Energy Commission at that time, recalls what occurred.

When a new bomb establishment was set up near Los Alamos, instead of asking the University of California to add this to its Los Alamos contract, I persuaded the Bell Telephone System to take this contractual responsibility.[12]

147

The scale of development and production functions that were soon located at Sandia is suggested by the fact that in 1963 it subcontracted with some 6,000 commercial and industrial suppliers from across the United States.[13] Sandia functions in much the same way as industrial organizations do in the development of complex technologies. A different illustration is the Fermi Laboratory, located near Chicago and established in 1968. Fermi was created to build and operate the most advanced high-energy physics research equipment then available. The contract for building and operating Fermi went to a new organization created by a consortium of universities. The use of a consortium gave control to the community of university physicists, who were both the facility's designers and its soon-to-be primary users. Thus the consortium arrangements gave responsibility for allocating scarce time on the research machine to those who were best able to assess the potential value of research. It was both technically and politically an efficacious arrangement.

The facility-management contract offers agencies a way to manage the synthetic innovation of large, complex systems. Such contracts allow for the rapid and flexible mobilization of both the physical facilities and the human resources needed to accomplish what is beyond the state of the art.

Prime Contract

Prime contracts with industrial organizations such as Lockheed or General Dynamics are the defense agencies' favorite vehicle for developing complex weapons systems such as aircraft, rockets, or ships. These contracts often symbolize the launching of or the modification of an organizational network focused on a specific weapons system. Prime contracts are critically important to the major defense contractors. First, they involve billions of dollars to be spent for the synthesis of weapons systems with superior or new capabilities. Second, they involve a nearly certain follow-on production-and-deployment contract worth many more billions. Thus prime contracts provide an assured source of income and profits for years into the future. Third, prime contracts have a powerful influence on (at the same time as they reflect) the future direction of national security policy. As such, they have a great influence on the general direction of technological innovation in the United States. Prime contracts are locomotives that pull whole arrays of technologies in one or another direction. The decision to issue a prime contract occurs only after a long period of research, conceptualization, and design. It represents the culmination of a consensus-making process within the defense complex that a superior or new innovation is possible and that its development can be carried out within a reasonable and predictable period of time. With rare exceptions, it is only after this consensus has been evolved that the Defense Department, NASA, or the Department of Energy go to Congress with a request for authorization to develop a technological system. When such consensus requests go to Congress, they are, in most cases, authorized. That authorization officially launches a large technological, economic, organizational, and political enterprise.

A development contract for a weapons system illustrates in microcosm the deterministic character of complex synthetic innovation. The objective of the contract and the role of the prime contractor is to develop a prototype of a technological system with performance capabilities that are beyond the state of the art. The word *system* is key to understanding prime contracts.[14] Modern airplanes, rockets, nuclear submarines, and tanks are complex systems, and the successful functioning of these systems requires an ability to integrate, synthesize, and achieve synergistic results from a large number of subsystems and components. The prime contractor is officially responsible for the integration and synergism of the components and subsystems that deliver the desired performance.

By the time a prime contract is let, the organizational network that carries out synthetic innovation has evolved a design on paper that will specify the desired performance capabilities of the system and the major subsystems and components. Defining these performance capabilities will be the result of years of continuously evolving scientific understanding and technological development. For both the system and the subsystems, the prime contract represents the best judgment of the defense-innovation complex concerning how far the state of the art can be pushed within a fixed period of time. This judgment is exceedingly complex and highly problematic. To specify too much performance is a formula for failure. On the other hand, to ask for too little performance is to face a future without the performance superiority needed to counter the perceived Soviet threat. Although there are many problems involved in developing new weapons systems, the greatest danger of having a weapons system development terminated occurs when performance standards that cannot be met are established. Indeed, the most powerful instrument of critics of weapons systems development is actual performance that does not meet contract performance criteria. A recent example is the Army's DIVAD gun, canceled after almost $2 billion in R&D expenditures.[15] Thus the process of establishing performance criteria is exceptionally important to the defense complex.[16] Those criteria always represent a compromise between what would be ideal and what is judged to be practically possible within a prescribed period of time. Design criteria for new weapons systems are an amalgam of the expertise within the organizational complex.

Besides serving as the integrators and synthesizers of subsystems and components necessary to a successful development process, prime contractors also play another role: they produce some portion of the subsystems and components. For example, airplane manufacturers such as Boeing and McDonnell Douglas generally build the airframe into which are integrated the many subsystems and components; most of these latter, however, are produced by subcontractors. The prime contractor, then, is the center of a complex organizational network consisting of thousands of subcontractors and suppliers.

Major development contracts involve pushing the state of the art of many of the subsystems and the overall system simultaneously. At both the system level and the subsystem level, this process of development occurs by continuous trial and error. Requirements for any subsystem are influenced by the performance of other subsystems. The synthetic process requires an exceptionally sensitive information system that links an ever-changing set of organizations that are both senders and receivers of information. For the innovation process to work, those who need information must receive it in a direct and efficacious manner. The prime contractor has a key responsibility to assure that information flow.

The Organizational Complex

Three events normally occur in the development of complex weapons systems: first, the initial performance criteria are continually adjusted and readjusted; second, some subsystems are unable to meet the performance standards initially specified; and third, some subsystems are able to develop performance capabilities that are far superior to those initially defined. Technological development and scientific understanding do not progress at a smooth, steady pace across a broad front. Major, unexpected breakthroughs can offer vastly improved performance capabilities and thus allow the establishment of much higher performance standards. Such breakthroughs are seen in the development of the transistor, the microchip, computers, and so on. Alternatively, technological development and scientific understanding may progress at a much slower rate than expected. The failure of a particular subsystem to meet expected performance capabilities can be dealt with in two ways. One is to modify (i.e., reduce) the overall performance requirements for the system. The other is to compensate for the lack of subsystem progress by substituting increased performance from other subsystems. For example, a common problem in aircraft design is weight. The inability to bring a subsystem within acceptable weight limits may be compensated for by developing other subsystems that weigh less. The impact that subsystems can have on development is further illustrated by the supersonic jet transport (SST). Boeing won the contract using a design that involved a "swing wing." A swing wing can be moved from a position nearly perpendicular to the fuselage at takeoff to a delta configuration for supersonic flight. As development progressed, Boeing could not produce a swing wing design within weight specifications. Increased weight meant that the passenger load had to be reduced, making the plane uneconomical. Failure to find a technical fix resulted in modified specifications that permitted Boeing to redesign the aircraft using a fixed wing similar to designs proposed originally—and unsuccessfully—by its competitors for the contract.[17]

Prime contracts, then, involve an extremely complex process of trial and error in multiple organizations, so the development process must be susceptible to continuous and rapid change. As we have seen, the organizational complex that carries out defense innovations has been able to build the appropriate organizational networks needed for each distinctive innovation. The key point is that the defense complex has generated organizational arrangements that have responded to substantive need as determined by trial and error. Where the goal requires exceptionally large and complex innovation, such as in NASA's Apollo Program, the needed organizational network may require creating a new federal agency. Where the objectives involve major incremental innovations, such as moving from one generation of aircraft to another, the network may require that an agency create a new organization, office, or group to provide oversight and organizational linking. A similar pattern of organizational creation or modification is likely to occur in a company receiving a prime contract. Not uncommonly, a Lockheed or McDonnell Douglas will create a new organization to carry out the development and later production of a new weapons system. This process of creating new organizations frequently has a domino effect on the whole subcontracting system that is part of the network. Driving the creation of new organizations is the need to tap and integrate the appropriate skills, capabilities, expertise, and the need to discard expertise that does not have utility.

The organizational networks established to carry out each innovation under the contract must ensure that those who need to exchange substantive information will be able to communicate directly. Information needs and substantive skills must govern organizational arrangements. Since needs and skills are continuously changing, organizational arrangements must be ad hoc. Although there is always an overlay of hierarchical organizational authority in government agencies and corporations, successful innovation precludes substantive exercise of that authority by individuals. Instead, success requires that decisions be evolved through the continuous exchange of information and the continuous process of experimentation carried on among those with the needed substantive skill and expertise.

On paper, a prime contract has a three-level hierarchy. At the top is the government agency; next is the prime contractor; and at the bottom are subcontractors. The reality is quite different. The agency, the prime contractor, and the subcontractors are intimate participants in a large and ever-changing team. In major development efforts it is seldom possible to identify exactly where decisions originated, how they were made, or what circumstances led to the decisions. Although there is a powerful need to maintain a paper trail—because of traditional notions of authority and responsibility—the paper trails of large development efforts are invariably misleading.

In their purest form, prime R&D contracts commit the agency to pay the direct costs, indirect costs (overhead or administrative), and a fee (profit). In essence such contracts say to the prime contractor, "Do what's necessary to develop a prototype, give us an accounting of your costs, and we'll send you a check." The prime contractor is the coordinator, integrator, and synthesizer of the work of thousands of subcontractors, many of whom are doing research and development. The prime contractor, then, may issue contracts that have the same cost-plus characteristic as the prime. In turn subcontractors may issue R&D—that is, cost-plus—contracts to subsubcontractors, and so on.

The prime contract is the epitome of a mechanism that allows government to buy new understanding and the capability to do what no one knows how to do for an explicit mission purpose. It provides the legal/managerial basis for building technology-specific organizational networks capable of synthetic innovation.

The Political Power of the Contract Federal System

The five contract forms that link together the organizations that make up the defense complex also provide the framework for a powerful political system. Quite clearly, the pattern of sustained support for defense technologies reflects their importance to the nation's security. But the political support for innovation and procurement of defense technologies also rests on a powerful political constituency that is sustained by the contract federal system. The complex links together a mélange of powerful organizations and individuals that together have demonstrated an almost irresistible political clout.

The basis for that linkage is the common interest of these diverse organizations and individuals in pushing the state of the art in science and technology that is relevant to defense innovation. Perhaps the clearest illustration of the effectiveness of this political system is seen in the context of prime contracts. The prime contract is a trigger; it provides the framework and sustaining support for a specialized political system. This system is built around an organizational network that can be utilized to ensure continued support for the development and procurement of the weapons system covered by the prime. Furthermore, the prime contract is the center of a procurement system that involves thousands of suppliers, subcontractors, and so forth. Indeed, any large prime contract will have suppliers and subcontractors located in all fifty states. It is a contracting system with subcontractors who have subcontractors who have subcontractors, ad infinitum. Tracing this network is almost impossible. That it is powerful is obvious.

As noted earlier, prime contractors and major subcontractors are likely to set up new organizations to carry out the major developments. The organizations and people involved in the innovation funded by the prime contract communicate with each other through a sensitive information exchange system, as we have seen. All participants understand that if the development is successful—that is, if it delivers successful prototypes—production contracts and deployment contracts will follow. Thus a prime development contract can be of great and immediate value to a large network of organizations scattered across the United States, and those organizations can expect long-term benefits (i.e., a secure future) when production of the new innovation begins. Not surprisingly, then, prime contracts and the networks built around them become active political networks. The slightest threat to a development project is rapidly communicated to those linked into the contracting network. Contractors, subcontractors, and suppliers respond immediately by contacting their congressional representatives. The political efficacy of the prime contract is impressive; it is able to mobilize broad-based support rapidly within the defense special-interest policy system.

The fact that agencies prefer to use private-sector prime contractors, rather than in-house laboratories or FFRDCs, is partly explained by this political clout. While FFRDCs and in-house laboratories can and do mobilize political support among their suppliers, such mobilization must be done with discretion and by indirection. The employees of in-house laboratories and FFRDCs are precluded from actively lobbying Congress. Private-sector prime contractors and their subcontractors have no such constraints, however. Large defense firms maintain in

The Organizational Complex

Washington their own lobbying organizations to monitor activities relevant to their well-being. Thus prime contractors actively use their subcontracting network as an information conduit and a system for political mobilization.

An illustration of this process was once recounted to me by a long-time observer of the contract federal system. He described the competition that developed between the Army and the Air Force in the 1950s over the responsibility for intermediate-range ballistic missiles. Both services had developed successful prototypes of these missiles, but the Air Force's missile had been developed by a prime contractor, while the Army's had been developed by its own facility, the Redstone Arsenal in Alabama.[18] In the end Congress gave operational responsibility for the intermediate-range ballistic missile to the Air Force. The man who told me this story said that the decision made sense given the Air Force's overall strategic nuclear responsibility. But, he noted, that was only one factor. When he had asked one senator why the choice in Congress was so overwhelmingly in favor of the Air Force, the senator answered, "The Air Force had forty-nine state delegations and the Army had one—Alabama."

The pervasiveness and effectiveness of prime contracts as frameworks for political support have made them very attractive to all of the federal agencies. The prime contract represents a clear example of the capacity of contract federalism to mobilize political support. In combination, the five contract types have been used by defense agencies to forge an organizational complex with a dazzling record of innovation. It is a system organized, energized, and managed in violation of the rules laid down by the secular trinity: it sets up a restricted (rather than free) market; it relies on teams operating in multiorganizational networks; and its goal is innovation (time) not efficiency (cost).

CHAPTER 9

Defense Contracting and the Secular Trinity

FROM THE END of World War II through the 1950s, the DOD and other defense agencies had great freedom in their R&D contracting. There were few congressionally imposed constraints on the agencies' ability to choose their contractors and negotiate sole-source, cost-plus contracts. In combination the Cold War, the recognition that defense was a unique government responsibility, and the central role of weapons systems innovation in providing for national defense allowed the defense organizational complex to develop with little more than rhetorical deference to efficiency and the free market. But as the costs of weapons systems grew and the pervasiveness of the defense contract federal system began to be recognized, the agencies' R&D contracting freedom began to be constrained. President Eisenhower's warning, in his farewell address, about "the military-industrial complex" was the symbolic beginning of this change. It was the Kennedy administration, however, that began actually to change the rules and restrict the use of sole-source, cost-plus contracts.[1]

A major effort was undertaken, especially in the Department of Defense under Secretary Robert McNamara, to "discipline" the contract federal system (i.e., to bring cost growth under control). The secular trinity guided that effort.

156

Defense Contracting and the Secular Trinity

Repeatedly, three types of events have triggered the efforts to change defense procurement. One has been the pattern of what are widely seen as gross overcharges, represented by $7,000 coffeepots. The second is the consistent pattern of each generation of military hardware costing some multiple of the preceding generation—that is, cost escalation. The third is the pattern of prototype costs far exceeding their original estimates—that is, cost overruns.

The stock response to rising costs has been to seek greater *efficiency* through tighter management, changed contract terms, and more competition. The secular trinity has thus provided a consistent explanation of both the source of the problems and the appropriate response. The effort to use the secular trinity to govern defense contracting since the time of the Kennedy administration has led to an ever-growing body of legislation and regulations that have become steadily more detailed. The search for ways to ensure that defense contracting meets the standards of the secular trinity has resulted in a defense procurement system now so complex and cumbersome that it is without defenders.

Prior to looking at the evolving character of R&D contract procurement and management, some observations on the triggers for the pursuit of greater efficiency are useful. One of the intriguing elements of the contract federal system is the attention received by a $7,000 coffeepot in the context of a system that is spending $40 billion per year for R&D. No one seeks to justify $7,000 coffeepots. However, such items represent a minuscule portion of the cost of major development contracts. The tendency of the media to focus on coffeepots as opposed to composite materials or complex electronics systems results more from the ability to understand coffeepots than from their financial significance. Coffeepots are apparently assumed to be an indicator of a more pervasive problem of incompetence or venality.

The implication that $7,000 coffeepots are a function of some combination of government incompetence and private-sector venality seems, on its face, questionable. Those involved in the management of billions of dollars in defense R&D expenditures, whether they work for the agencies or for the contractors, are sensitive to the public relations costs of a $7,000 coffeepot. Anyone seriously interested in bilking the government in a system involving billions of dollars of exceedingly complex activities would surely do it in areas unintelligible to auditors and laymen. The likely explanation for the $7,000 coffeepot is that it reflects the processes required to achieve synthetic innovation. Development contracts start by establishing performance capabilities that are beyond the state of the art. R&D decision making occurs within a complex, ever-changing organizational network and is driven and informed by trial-and-error activities. The high performance standards established for the overall system inevitably cascade down to the subsystems and smallest components, imposing on them unique design requirements. Unique designs of anything are expensive to produce—especially in small numbers of items. Synthetic innovation attempts to achieve performance that is beyond the state of the art by pushing subsystems and components. The assumption is that subsystems and components are critical to the performance of the system.

Where decision making occurs in an ad hoc fashion—that is, where it occurs by consensus recommendations moving from group to group—it is not surprising that coffeepots end up having unnecessary performance requirements. In fact, it is to be expected that a system of contracting that starts with the assumption that superior performance is critical to national defense will inevitably build superior performance into subsystems that don't need it—even coffeepots. The critical problems spotlighted by $7,000 coffeepots and the media and congressional attention they receive are likely not venality and incompetence but rather an illustration that we live in a society that fundamentally misunderstands the process whereby synthetic innovation occurs.

Within the defense special-interest policy system, concern with the innovation process has been driven much harder by the rapidly rising costs of each new generation of hardware and the consistent pattern of cost overruns associated with major hardware development contracts. Many of those involved in the defense innovation system believe that these costs could be significantly reduced through better management. A frequently used illustration of the rapid escalation of weapons systems costs is fighter aircraft. It is regularly noted that the World War II P-51 fighter cost roughly $100,000, the Korean War F-86 cost roughly $1 million, and the Vietnam War F-4 Phantom cost roughly $5 million. The cost of today's F-16 is in the range of $15 million to $20 million per copy. Clearly, developing superior and/or additional performance capabilities has driven the costs of military technology along a much steeper curve than is reflected in general inflation. Cost overruns on the development of major weapons systems are the norm. In the early 1960s, a study by Peck and Scherer found that for twelve weapons systems, on average, real costs were 3.2 times higher than estimated costs.[2] The same pattern appears to have held for the last twenty-five years.

One of the most intriguing aspects of the defense contract federal system is the degree of indignation expressed by those involved in the system concerning both cost overruns and rapid escalation in the cost of new weapons systems. Those who purport to manage the system frequently reflect the same degree of ignorance of what is critical to successful continuous innovation as do laymen. Thus it is regularly the case that both laymen and those responsible for managing defense R&D share a common response to each scandal or new evidence of the rapidly rising costs of defense R&D. They respond with new legislative mandates and administrative regulations that apply more detailed and more stringent terms to R&D contracts and to how R&D contractors are selected and managed. These efforts to legislate and regulate with the goal of achieving efficient innovation have had two major foci. The first involves designing contract terms that create incentives for contractors to be more efficient and that allow for detailed management (micromanagement). The second involves pursuing efficiency by increasing the competitiveness of the contract allocation process.

Contract Terms

The range of experiments with contract terms designed to deliver efficiency is large.[3] In principle, however, all experiments seek to do the same thing: they seek to deliver the benefits of a free market—that is, to make efficiency rewarding and inefficiency costly to the contractor. The vigorous pursuit of efficiency-producing contract terms began during the 1960s when the Department of Defense sought to stimulate efficiency by using the firm fixed-price contract (FFP). The FFP included contract terms that were at the opposite pole from the cost-plus-fixed-fee (CPFF) contract, which had dominated R&D contracting from the end of World War II until the election of President Kennedy. The goal of the FFP was to place the financial burden of inefficiency on the contractor instead of on the government. Robert Art describes this process in his study of the development of the TFX, later F-111, airplane.

> Under this contract the seller would agree to develop a weapons system at a price that, once negotiated, would remain fixed. The price could not be revised upward (or downward) to reflect changes in the seller's estimates of his costs or his experience in incurring them. Thus, under the FFP, because the price remained fixed, the profit would be the residual of price less cost. Because costs and profits were inversely related, the seller would have a strong contractual incentive to keep his costs as low as possible. In this way he could maximize his profits. For the same reason he would have a strong contractual incentive to estimate his costs as accurately as possible. If he underestimated them and consequently incurred greater costs than planned then he and not the government would be the one to pay them. His profits would correspondingly be reduced.[4]

The FFP contract, on its face, builds in powerful incentives for the contractor to operate efficiently, since efficiency results in increased profits. There is one flaw in this approach, however. Efficiency is a measure of how to do something for less cost, but to measure relative cost requires that you know how to do it and already have a cost standard. No such standard exists when the contract calls for doing something that has never been done before and that no one knows how to do.

Competitive Bidding

Secretary McNamara also sought efficiency through competitively bidding contracts. Here a central concern was the inefficiency presumably reflected in cost overruns. Both contractors and federal agencies have traditionally had incentives to use low initial estimates for large development contracts. Low estimates facilitate congressional authorization. Since no one knows what the costs are going to be and how they will be spread over many years, there is little point in using high estimates. Precise cost estimates are simply not possible in the absence of detailed specifications for what is being purchased. The best that can normally be done is to use historical records and estimate a multiple of the cost of the last development. Where there is no past record, as is the case with the Strategic Defense Initiative, any cost estimate is suspect—at best, just an educated guess.

The search for efficient innovation through the use of marketplace competition runs up against the problem of how one chooses among bids to do the unknown. The competitive process usually begins with an agency's requesting design bids, which describe in general terms how the bidder will do something no one knows how to do and at what cost. The agency must then seek to choose the most cost-effective approach. The choice requires judging both the probability that the contractor will produce the desired synthesis and the likelihood that the proposed costs are realistic. This must be done in the absence of either the agency's or the bidder's understanding how the unknown will be accomplished.

Something of the difficulties involved in competitive R&D contracts is suggested by Robert Art's study of the TFX. Art describes the TFX (later the F-111) contracting process as moving through several phases.[5] It began with independent efforts by the Air Force and Navy to formulate initial performance criteria for two aircraft with distinctively different missions. Under pressure from McNamara, Phase Two saw the TFX concept created as the two services merged their performance criteria into a single biservice aircraft capable of carrying out multiple missions. The desired performance criteria called for both a land-based and carrier-based aircraft that was a bomber, a fighter, a reconnaissance aircraft, and a transoceanic airplane capable of delivering either nuclear or nonnuclear weapons. Phase Two produced a set of performance criteria for an aircraft the aircraft industry was asked to bid on. Six companies submitted preliminary proposals; the Department of Defense selected two of those proposals and awarded study contracts to Boeing and to a General Dynamics–Grumman team. Phase Three involved these two teams developing ever more detailed designs on paper. Specifically, each team went through three designs; the final one was the basis for selecting the prime contractor. The Navy and the Air Force recommended the Boeing design, but McNamara chose the General Dynamics–Grumman team. Not only did the two services prefer the Boeing design, but the Boeing design also had a lower estimated cost.

McNamara's decision triggered vigorous criticism in Congress, ranging from questions about technical design and performance to disputes over where the contract selection decisions should be made (in the services or the civilian secretary's office) to pork barrel charges. The McNamara decision placed development of the TFX in Fort Worth, in Texas, the home state of Lyndon Johnson, then vice-president.[6] A Boeing decision would have placed the contract in the state of Washington. Hearings on the TFX decision continued in the Senate Permanent Investigating Subcommittee for almost ten months. McNamara justified his decision by claiming that the General Dynamics–Grumman proposal was more cost-effective, while critics of the McNamara decision argued just the opposite. In the end the TFX development remained with General Dynamics–Grumman.

The TFX controversy is particularly instructive because it reflects the impossibility of meaningful cost-effectiveness assessments between proposals aimed at synthetic innovations that require pushing the state of the art a great distance. The TFX performance requirements were vastly different from and greater than those of any aircraft ever previously delivered. Robert Art characterizes the situation as follows:

> Each contractor had only put designs on paper, not flown planes in the sky. Each group [McNamara versus the Air Force and

Navy] was therefore judging proposals of both contractors, not the result of their development efforts. But because the uncertainties are greater at the design stage than at the development stage, even the technical experts disagreed on both the probability and desirability of attaining the design promises. Because there could be no empirical proof at this stage in development to validate or invalidate either group's judgments, both remained valid.[7]

The experience with the TFX competitive contracting highlights the conflict between existing ideology and reality. Objective, empirical, valid, reliable efficiency judgments are not possible when one is contracting to do something no one knows how to do.

Consequences of the Pursuit of Efficiency

There is perhaps only one observation that can be made with absolute certainty about the continuing effort over the last twenty-five years to find ways to achieve efficiency in the R&D contract process. No record of empirical evidence exists to support the notion that one set of contract terms versus another results in greater R&D efficiency. Similarly, there are no hard data to indicate that competitive bidding for R&D contracts has led to efficiency. The other side of the coin is that there is no empirical evidence to indicate that these efforts to achieve R&D efficiency have *not* generated efficiency.

In the absence of convincing evidence, one point is clear. Those responsible for writing legislation and establishing regulations governing R&D contracting continue to subscribe to the ideological faith of the secular trinity. Year after year, as cost overruns continue, as the cost of new weapons systems escalates, and as purportedly convincing evidence of the deficiencies of the system are demonstrated by $7,000 coffeepots, more stringent legislation and more detailed regulations to require efficiency and good management are put in place. The number of managers, lawyers, and auditors who now function within the defense contract system has been growing exponentially. As a consequence of all this, the system has become more and more cumbersome. It has changed from being one that was essentially free of litigation to one that is now permanently embedded in it.

As efficiency-driven regulations and legislation grow, the defense organizational complex has had to find new ways of manipulating the legislation and regulations to ensure that the incentives for experimentation are maintained. The continuing reality is that the financial risks of experimentation to R&D organizations must be limited, and the performance criteria of both systems and subsystems must be perpetually adjusted to respond to the changing capabilities of technologies. With each effort to write more detailed requirements comes the need to make exceptions to the requirements. Every change in system or subsystem specifications and requirements now requires that someone from the contracting agency sign off on a change order. That is normally not too difficult because the people from the contracting organization, if they are to be effective, must be part of a team whose expertise is evolving by means of the trial-and-error process. The result is a huge, complex (if misleading) paper trail. Whether costs are reduced when such complexity is entailed is an open question.

The pursuit of efficiency in extreme cases has led prime contractors to the point of bankruptcy. The classic case is the Lockheed Corporation. Lockheed signed a total package contract to develop the C-5A cargo plane, a plane requiring performance far beyond the state of the art. The total package contract involved the Air Force's committing a fixed amount of money to cover both development and production of a fixed number of planes. Lockheed was unable to meet the production terms with the funds provided. Some critics argued that Lockheed should have been required to meet contract terms out of its own assets, but there was one unfortunate barrier: meeting the contract would have bankrupted the corporation. In the end the Air Force covered a substantial portion of the costs of producing the C-5As.[8] The reasons were simple. Lockheed is a prime player in the defense organizational complex that carries out continuous innovation and was involved in development and production of numerous other weapons systems. Dissolving Lockheed would have required building another organization like it to take its place. In so doing, a substantial amount of its experiential learning would have been lost and a large number of ongoing programs dealing with other weapons systems would have been severely disrupted. The reality is that there was never any chance that the Air Force would have forced Lockheed into dissolution.

The same situation occurred when poor management and dishonesty were identified within General Dynamics. The prevailing ideology says that the way you discipline General Dynamics is to withhold payments for existing work or to deny General Dynamics the opportunity to bid on future contracts. The great dilemma is that, if you believe the Trident submarine, the F-16 fighter, and the M-1 tank are critical to the nation's national security, it is extremely difficult to take severe and damaging sanctions against General Dynamics since it is the only organization making those three weapons systems. The threat of nonpayment or denial of bidding opportunities on future contracts may, in fact, be symbolically important, but they are, in the short term, fraudulent threats. Like Lockheed, General Dynamics is integral to the successful process of U.S. defense innovation.

In truth, the General Dynamics problems and the opportunities for dishonesty may come in part from the efforts to apply regulations and standards that require efficiency. In a company that has multiple R&D contracts but that is attempting to operate within firm fixed-price contracts, there is a powerful temptation to charge people, materials, and time against other contracts that have available money. The process of continually trying to do what no one understands how to do, while applying rigid cost standards, makes it attractive for organizations that depend upon maintaining a state-of-the-art capability to manipulate funding so as to maintain that capability. In truth, such manipulation may be essential to national security.

Finally, it is quite possible that the whole effort to improve efficiency may be inhibiting innovation.[9] One of the first tendencies in efforts to improve an organization's efficiency is to bring in people to monitor it. All too frequently, those people are accountants, lawyers, and managers. The growing procurement bureaucracy is believed by some close observers to be a major cause of slowdown in innovation. One result of this problem has been a tendency on the part of the military to give development projects a high-level security classification. Certainly one purpose is to secure sensitive information, but an increasingly important benefit is that classification allows the DOD procurement bureaucracy to be partially circumvented. The highly classified (black) programs allow for more integrated and more rapid development.[10] Thus the pursuit of efficiency may be inhibiting innovation while at the same time there is no evidence that efficiency (i.e., lower cost) is being accomplished.

Four decades after the end of World War II, two things stand out. First, a unique organizational system has delivered continuous synthetic innovation in defense. Second, for the past three decades there have been continuing efforts to force that system to meet efficiency standards in a situation where efficiency cannot be measured. The upshot has been bigger and bigger bureaucracies and longer and wider paper trails. Today efforts to apply stringent legislation and regulation must be inventively subverted if innovation is to be achieved. Both the pursuit of efficiency and its subversion involve more and more people, more and more time, and probably represent an increasing impediment to innovation. Efforts to apply the standards of the secular trinity to the defense innovation system have done little to alter the pattern of cost overruns and escalating weapons systems costs. The critical answer to the management of defense innovation is not in the pursuit of efficiency. Rather, the critical choices concern how many and which major systems should be developed. The reality is that the rising costs of weapons systems flow from a combination of the growing menu of options and their increasing complexity. Developing a large number of options not only involves high development costs but also results in limited individual production runs. Limited production runs preclude economies of scale and thus mean a high unit cost for each system.

The synthetic society does not allow this fundamental challenge to be sidestepped through so-called "improved management," reflected in the manipulation of contract terms and increased competitive bidding aimed at efficiency. The development during and after World War II of a contracting system that allowed agencies to judge which organizations offered the most likely prospect of pushing the state of the art and that allowed for sole-source, cost-plus contracts reflected the reality of the synthetic society. The evolution of the defense procurement system since 1960 reflects an effort to force reality to conform to ideology. In the end that effort must fail.

Both a reenergized innovation capability in defense and a meaningful response to the export race require fundamental changes in public policy, not the pursuit of efficiency.

CHAPTER 10

Medicine and Agriculture

THE SIZE, cost, diversity, and complexity of defense innovation and the organizational complex that carries it out are much greater than are the corresponding elements in medicine and agriculture. Fundamentally, however, synthetic innovation in defense, medicine, and agriculture is the same. In all three sectors, government covers a large portion of the front-end (R&D) costs, serves to link public- and private-sector organizations into a complex capable of creating organizational networks, and acts to create market incentives for innovations. In each of the sectors, special-interest policy systems exist that have innovation as a policy goal. And in each case the special-interest policy systems have the necessary size and strength to provide a fertile environment for the synthetic organizational complexes.

It is, however, useful to look briefly at how the special-interest policy systems and organizational complexes in medicine and agriculture compare with defense. What follows is a broad grid map of the origins and arrangements that provide for continuous innovation in medicine and agriculture. This map emphasizes two points: first, in each of the systems, close government/private sector interaction is essential to success; second, the particular roles and relationships between the public- and private-sector organizations vary, depending upon their historical origins and the substantive character of their innovations.

Medicine

The roots of the special-interest policy system for medicine and the role of science and technology in defining problems and identifying policy options run deep in American history. Harvard's Don Price provides perspective as follows:

> Throughout the nineteenth century judges were gradually persuaded that public health regulations were necessary for the prevention of epidemics. These regulations were extended from the simple provisions of quarantine to a wide variety of sanitary regulations, and finally to the provision of authority for zoning and city planning in the interests of the public health and safety. Gradually the massing of medical evidence regarding contagious disease, and the relation of such evidence to the sociological statistics regarding urban life in general, provided the basis for the development of municipal powers in America.[1]

What began in the last century as a series of ad hoc local responses by courts to the immediate threat of epidemics has since come a great distance. Most Americans today assume that good health care is a right. Government is expected to ensure that such care is available and in the last analysis to provide it for those who are otherwise unable to avail themselves of it. In sum, Americans see health care as a distinct governmental function. With the development of these public attitudes the president and Congress have built and steadily enlarged the special-interest policy system for medicine.[2] An increasingly high-priority goal of that policy system is to increase scientific understanding and the techniques, technologies, and capabilities necessary to prevent and cure medical problems. An ever-changing menu of health threats has provided a rationale for continuous innovation similar to that which the Soviet threat has provided for defense.

An illustrative map of some of the key participants in the special-interest policy system for medicine looks as follows. First, at the federal executive level, there is the Department of Health and Human Services (HHS). Within Congress there are committees in both houses primarily concerned with medical care. In addition, each of the states has one or more departments concerned with health. Then, in the private sector, participants range from pharmaceutical companies, hospitals, doctors, and insurance companies through a wide range of broad-based, organized interest groups. Finally, in the nonprofit research sector, there is a large collection of universities and teaching hospitals. The special-interest policy system for medicine has policy jurisdiction over a set of activities that together account for 11 percent of the nation's GNP. It is a policy system predisposed to technical solutions to problems; thus it is powerfully supportive of the organizational complex that delivers continuous synthetic innovation.

Although the role of government in medical innovation is major, it is much less centralized than in defense. For example, DOD funds R&D, ties the public/private-sector organizations together in a contract federal system, and buys the innovations. In medicine, however, different organizations play different roles.

The key role in R&D funding is played by the National Institutes of Health (NIH).[3] Although the NIH is heavily oriented toward research (i.e., developing understanding), it is divided organizationally into individual institutes that are focused on distinctive categories of medical problems. Perhaps the best known is the Cancer Institute, which is the major NIH participant in an organizational network focused on cancer. In 1988 the NIH spent over $5 billion on medical R&D.[4] A major portion of this NIH money goes to researchers in universities and teaching hospitals, in the form of competitive grants and contracts,[5] with the other large portion going to support the in-house R&D capability of NIH, located in Bethesda, Maryland. Most external NIH support is provided through grants. Little NIH funding involves anything equivalent to the prime contracts in defense, which commit the performer to delivering predetermined capabilities within fixed time periods. In sum, NIH funding arrangements reflect the substantively different character of medical innovation. Medical innovations that address major problems (e.g., polio, AIDS) tend to be radical in character—that is, they provide a capability that is new and is based on research discoveries. The goal is a cure for the medical threat. Defense innovation, by comparison, is generally incremental in character, reflecting the acceptance that there is no permanent technical cure for the Soviet threat.

The development and testing of technologies and techniques in the organizational complex for medicine tends to be carried out in industry, universities, teaching hospitals, private and nonprofit laboratories, and by medical practitioners. The linkage between the university/teaching hospital community and industry is broad-based and complex, and government's role in facilitating and managing this linkage is limited. In the cases of some technologies and treatment techniques, however, the NIH does fund and facilitate development and testing. In other cases industry, especially pharmaceutical companies, develops drugs and funds their tests. In some instances the initiative and development comes from researchers and practitioners in the universities and teaching hospitals. Finally, foundations and organizations such as the American Cancer Society support and promote development and testing of some innovative technologies and techniques.

The organizational complex that carries out synthetic innovation in medicine enjoys an environment that provides a nearly guaranteed market.[6] This market is furnished by an insurance system that will pay for the use of any new technology determined by the medical community to be efficacious. Two points warrant emphasis. First, insurance companies exercise few cost criteria in providing coverage for new drugs, technologies, and treatment techniques. Second, neither the medical community nor the Food and Drug Administration use cost criteria in their efficacy determinations. The efficacy judgment rests on whether the technology or technique works as a treatment or preventative for medical problems. As in defense innovation, the measure of success is performance, not cost.

In general, efficacy decisions are made in two ways. For new drugs the determination rests with the federal Food and Drug Administration, which must officially approve their use before they may be legally sold. For most other technologies and techniques efficacy is determined by a predominantly informal peer-approval process. This may occur through papers presented at professional meetings or through peer-reviewed articles that report the performance of innovative techniques and technologies. Efficacy may be determined also by NIH cooperative studies conducted essentially by convening groups of experts who review the data on the techniques or technologies.

Decision making on the use of innovative technologies and techniques rests primarily with the community of medical practitioners. Thus market success rests neither with those on whom the innovations are used (patients) nor, for the most part, with those who pay for their use (the insurance programs and/or the patients). For this reason, marketing primarily involves communicating the efficacy of new techniques and technologies to a practitioner community that has powerful incentives to use medical innovations offering improved results. Such communication occurs in several ways: (1) through medical journals and other professional media; (2) through workshops and seminars; (3) through representatives for products (e.g., pharmaceutical company representatives); and (4) through informal (e.g., word of mouth) communication.

A brief look at the special-interest policy system for medicine and the organizational complex that produces continuous synthetic innovation highlights several points. The policy system and the organizational complex did not have their origins in a major triggering event, as the defense system had its origins in the single crisis event of World War II. Rather, the system grew incrementally in response to a diverse range of medical threats. As with defense, it is widely believed that government has a central role in medicine because good health care is seen as a basic human right. This belief provides a rationale for not conforming to the strictures of the secular trinity.

The powerful special-interest policy system for medicine operates on the assumption that continuous innovation is a major instrument for solving existing medical problems as well as a necessary instrument for responding to future problems. The system therefore ensures stable support for continuous innovation. In this system government plays major roles in what economists call the "technology-push" and the "demand-pull" functions in medical innovation, but it plays a less central role than in defense in linking together the institutions responsible for moving ideas through development into use. In medicine part of this linking role is played by a diverse collection of professional, organizational, and economic incentives provided by nongovernmental institutions. Through its more than $5 billion funding of medical research and development, NIH plays the key technology-push role. Medicare, through its support for medical services, provides a powerful demand-pull for medical innovation. Medicare's influence is suggested by the fact that it finances health care for thirty-one million Americans and provides 40 percent of all hospital revenues derived from treating patients.[7] In fact, however, Medicare's role is probably greater, in that it also sets the framework and expectations for private-sector insurance programs.

That medical innovation in the United States sets the world standard is beyond question. Medicine, like defense, represents a clear synthetic innovation success story.

Agriculture

The origins of government-supported and -directed efforts aimed explicitly at innovation in agriculture go further back in American history than do those of any other sector. The beginnings of the special-interest policy system for agriculture and the precursors of the synthetic organizational complex can be traced to the nation's origins. Again, Don Price offers a perspective:

First of all there is the story of the agricultural sciences. In this field government support was no novelty, even at the time when President Washington first proposed to the Congress the establishment of a National Board of Agriculture. Parliament under the Puritan Commonwealth had granted funds for experimentation in Georgia on the growth of indigo and other agricultural products.

The creation of a federal agricultural agency, like the rest of the grand design of the Federalists for the development of the national economy, was blocked by Jefferson's strict construction of the Constitution. Nevertheless, Jefferson as an individual gave a great impetus to the support of the agricultural sciences. In 1787, for example, he smuggled rice out of Piedmont in spite of the laws prohibiting its export on pain of death, and encouraged the found-ing of agricultural societies and their cooperation on a national basis.[8]

In contrast to defense and medicine, government support for and involvement in agriculture cannot be rationalized in terms of a noneco-nomic, uniquely governmental function. Rather, the most common rationale is the need to protect and sustain the "family farm." Within the mythology of American society, the family farm appears to call forth images of special virtue. In the Jeffersonian formulation, the yeoman farmer was the rock upon which a democratic system would be built.[9] The resilience of that imagery over two hundred years is impressive. In truth, however, while the imagery of the family farm has been useful, the growth and power of the special-interest policy system for agriculture and the organizational complex that has delivered con-tinuous synthetic innovation have been driven by a much more mun-dane engine. That has been the political clout of agriculture, especially in Congress. Agriculture's clout has been clear and continuing for two hundred years, and there is no evidence of its early demise. Similarly, the evolution and growth of agriculture's synthetic organizational com-plex has been continuous for over one hundred years.

The special-interest policy system that is the present manifestation of this long historical evolution is large and sophisticated and has continuous synthetic innovation as a prime goal. It has grown in re-sponse to a series of economic, natural, and disease crises that have led to the repeated search for technological fixes—thus the rationale for continuous innovation.

A brief map of some of the key participants in the special-interest policy system for agriculture looks as follows. At the federal executive level the primary player is the Department of Agriculture (DOA). Within Congress there is an agriculture committee in each house. The states have departments of agriculture and a wide range of other units that focus on specific products—wheat, for example. In the private sector are large numbers of farmers, companies that supply and/or buy and process the output of the nation's farms, as well as large numbers of organized interest groups (e.g., the American Farm Bureau Federa-tion). Finally, there are a large number of R&D organizations, the most prominent being agricultural colleges and experiment stations associated with the land-grant universities.

The special-interest policy system for agriculture has the dominant role in making and modifying policy for the nation's food-production system, which accounts for nearly 10 percent of the GNP.[10] The key role played by technology in the growth and development of American agriculture makes the special-interest policy system powerfully supportive of the complex that delivers continuous synthetic innovation.

Agricultural innovation has rested heavily on DOA, which has exercised the same kind of centralized role represented by DOD in defense. Alternatively, the instruments used by DOA and its degree of control are very different than is the case with DOD. The two most distinctive differences are reflected in the way R&D funding is handled and in the way market incentives are created. Specifically, little R&D is contracted out and the market incentives for innovation provided by government are both fragmented and vary over time.[11] The vast majority of the R&D expenditure by DOA, over $1.1 billion in 1988, goes to its own research facilities or to the land-grant universities.[12] The money for R&D support at the land-grant schools is primarily allocated on a formula basis rather than through competitive contracts and grants, as is the case in both defense and medicine. Simply put, most agricultural R&D is allocated as part of a continuing line-item budget process.

This approach to R&D reflects the early, more conventionally political origins and evolution of the synthetic innovation complex for agriculture. Its origins are in the Morrill Act, passed in 1862 as part of the effort by the new Republican Party to build a political base among farmers. The Morrill Act granted the states federal lands to support the creation of what are now known as the land-grant colleges for the purpose of providing training in the agricultural and mechanical arts. By the beginning of the twentieth century, federal cash grants were being provided to support R&D in experiment stations associated with the land-grant colleges. From the point of view of the research community, this was not the best way to gain high-quality research. Those in the traditional universities communicated their views by condescendingly referring to the land-grant schools as "cow colleges." In fact, however, the Morrill Act created a nationwide network of applied research and teaching institutions that have generated worldwide envy. In their focus on the practical problems of local farmers, the land-grant schools have been key contributors to the incredible productivity of American agriculture.

Given the overt political needs that launched the land-grant colleges and the pattern of political pressure that has sustained their federal support, it is understandable that R&D money has been allocated differently than in defense and medicine. Given the unique historical origins of agricultural R&D, it is not surprising that funds are divided among the states on a statutory (pork barrel) basis, not a discretionary administrative or scientific merit basis.[13]

Government R&D support for agriculture is not distributed to a wide range of nongovernment organizations through competitive grants and contracts, as it is in defense. As in defense, however, government supports the full range of activities, from research through the development and testing of new products and techniques necessary for innovation.[14] Further, with government support the innovation complex for agriculture developed a broad-based system to facilitate diffusion (marketing) of technologies—i.e., the county agent system.[15] County agents are employees of the colleges of agriculture. Their original functions were to communicate to farmers the correct use of the technologies developed in the agricultural colleges and to communicate to the colleges the problems that farmers needed solved. The innovations developed with DOA support were from the beginning carried out with the explicit purpose of their being transferred to private-sector organizations and individuals. Utilizing technologies and techniques developed within the DOA-supported innovation complex, a large private-sector industry developed to supply the nation's farmers and to process and sell their products. Today that industry ranges from seed and chemical companies through machinery manufacturers to companies that produce synthetic food products such as those derived from textured vegetable protein.

The agricultural supply and processing industry is tightly linked to government-supported R&D organizations. Over the years the industry has added its own large-scale R&D capabilities, and its marketing system has substantially replaced the county agents as the vehicle for diffusing new and superior-performing technologies. In combination, the private-sector marketing system and the increasing receptiveness of farmers to new technologies, which has resulted from experience and socialization, has created a system that assures rapid utilization of innovations. The record of success in moving new ideas and methods to rapid use has led many observers to conclude that there is no longer any need to maintain the county agent system. A significant element in the speed with which new agricultural technologies are utilized has been a set of agricultural programs that have created incentives for using agricultural innovations. For the most part, government programs aimed at manipulating the market for agricultural commodities have been triggered by problems and crises caused by nature (e.g., drought), disease or pests, and/or overproduction. The general pattern of agricultural programs has been to put a floor under farm income in the short term and to find a technological fix to take care of the longer term.

One response to natural disasters such as drought has been development of new practices such as no till farming, to protect against loss of moisture and wind erosion. The generic pattern has been government programs that make the receipt of cash payments or other subsidies contingent upon the farmer's using new technologies and techniques. In the case of disease and pests (e.g., boll weevils), the synthetic innovation complex has sought understanding and technological fixes, the utilization of which were then subsidized by other agricultural programs such as irrigation programs for cotton growing in dry climates. The programs dealing with the chronic problem of overproduction have had two characteristics. One is to provide mechanisms or incentives for limiting production or sale of the surplus commodity. In the case of oranges, the approach is to give cartel control to producers so that they can limit sales and thus maintain the price at a profitable level. In the case of corn, the pattern is to guarantee a floor price to the farmer in exchange for his taking some portion of his land out of production.

In all cases, however, the programs have as their long-term goal pushing farmers to become more efficient. This goal of efficiency has been pursued with programs aimed at increasing productivity; this has been achieved by increasing the productivity of both the farmer and the land. Use of improved seeds, fertilizer, pesticides and herbicides, improvements in farming practices, the development of larger, more efficient machinery, and the use of larger farms has resulted in steadily increasing productivity. Increased productivity has led to a continuous cycle of overproduction, which has in turn led to even greater efforts to increase productivity.

With each evolution of farm programs that provide floor prices for crops in exchange for taking land out of production, farmers have withdrawn their least productive land. Then they have used the most advanced technology to maximize the productivity of their best land. In sum, farm programs have stimulated innovation by providing economic incentives for utilizing the highest-performance technologies. Such action has in turn exacerbated the overproduction problem.

Allocation Problems

As in defense, synthetic innovation in medicine and agriculture has been a continuing success story. With that success has come an ever-growing demand for the nation's resources. In the case of medicine, the resource demands are reflected both in the cost to government medical insurance programs and in the portion of our GNP devoted to medicine—roughly 11 percent in 1988. For agriculture, the overall cost of food has remained low, but the cost of government agricultural subsidy programs grew during the 1980s from \$2.8 billion in 1980 to \$23.1 billion in 1987.[16] Faced with rising costs, the policy/administrative response has been similar to that triggered by rising costs in defense, namely, a set of actions derived from the secular trinity. The goal has been to achieve efficiencies (lower costs) through better management and greater use of the free market. Specifically, the response has been to increase competition and provide incentives (i.e., larger profits) for individuals and organizations to be more efficient.

In the case of medicine, the approach derived from the secular trinity has been to increase the costs paid by patients, to provide review of diagnostic and treatment techniques used by doctors and hospitals, to move hospital care from nonprofit to profit-making hospitals, and to increase competition among those delivering medical services. Perhaps the ultimate example of the faith in private-sector efficiency has been the recent proposal to turn the in-house R&D capabilities of the NIH over to the private sector.[17] Whatever may be the relative benefits of these actions, one point is clear: medical costs continue to rise at a much higher rate than the economy's general inflation growth. Thus medical costs are absorbing an ever-growing share of the nation's resources. Improved medical care is keeping people alive longer, and the medical costs of people increase with age. In parallel, ever more sophisticated medical technology and techniques are more and more expensive. In sum, the ability of technology to keep people alive generates a growing demand for new diagnostic and treatment capabilities that cost more. The increasing resource demands of medicine, which result from the deterministic character of the special-interest policy system for medicine, and from its organizational complex, which delivers continuous synthetic innovation, cannot be effectively understood and managed using the secular trinity.

In the case of agriculture, the problem has the same generic character but functions in a different setting. The generic problem is that the process of continuous innovation results in ever-improving productivity, which leads to overproduction and low prices. A standard analysis is that this problem is the result of the long history of government intervention in and disruption of the market. Thus there are repeated proposals for government to withdraw from agriculture. These proposals are based on a belief that the farmer, left to his own devices in a free market, would solve the problem.

The power of the special-interest policy system for agriculture repeatedly precludes adoption of the free-market approach. The results are farm programs that are jerry-built compromises rationalized as transitory responses to immediate crises. The more general belief—more accurately, the illusion—is that such programs will be unnecessary when the economic nirvana of a free market arrives in the future. However, two developments preclude the arrival of that happy state. First, the accumulated successes of technological innovation in agriculture have given the world a capacity to produce surplus food. In the synthetic reality there is a permanent world capability to overproduce in agriculture. Only a very small number of countries, like Japan, which have large populations relative to arable land, appear permanently condemned to food-production deficiencies. In most instances where national food-production deficiencies exist, they result from the lack of the necessary economic/technological infrastructure or from social/ political instability. Once technology made agriculture a science-based as opposed to a resource-based sector, overproduction became the permanent problem. The second development that fundamentally changed the agricultural problem also has its roots in technology. It is that modern transport technologies have created an international agricultural market. But it is not a free market. To the contrary, almost without exception governments actively intervene in the market either to protect their agricultural sector or to gain needed export earnings.

Intervention aimed at the protection of the agricultural sector is driven by two concerns. One is the political power of farmers. Farmers demand that their governments provide economic protection, and they become a source of political instability when their economic well-being is threatened. A second drive is the desire of nations to be agriculturally self-sufficient or at least as self-sufficient as possible. Countries that have experienced food shortages nearly universally strive to prevent that happening again.

Thus food-producing countries with high costs protect their agriculture against low-cost producers with tariff barriers and subsidies. Faced with surplus agricultural commodities, many countries with high-cost production subsidize exports as a way to reduce the surpluses, a pattern followed by the European countries. Therefore, not only is U.S. food kept out, but high-cost national production becomes a subsidized source of competition in the world market. Some countries also subsidize exports because of their need for export earnings. Countries with negative trade balances and large agricultural surpluses find it advantageous to sell on the international market at prices below production costs when that provides a way (sometimes the only way) to service debt and pay for needed imports.

Because of worldwide use of advanced agricultural technologies, the surpluses they provide, and pervasive government intervention in the agricultural economy, there is no free-market answer to the U.S. problem of overproduction. Put simply, efficient (low-cost) production does not translate into a competitive advantage for the American farmer.

Farmers cannot compete with governments and win in the export market.

As in defense and medicine, the problems of agriculture in the synthetic society cannot be accurately understood and addressed using the secular trinity. In all three sectors the capacity to deliver continuous synthetic innovation has produced technological leadership for the United States. In all three sectors the United States is competitive in the export market. Moreover, in all three sectors the resource demands created by continuous innovation are absorbing a growing share of the government's and/or society's resources.

Success (i.e., leadership in continuous innovation) rests on the existence of large, well-established special-interest policy systems to support organizational complexes that deliver synthetic innovation. Government's role in synthetic innovation in defense, medicine, and agriculture is a threefold one. Government pays for high-cost front-end R&D, participates in linking diverse public- and private-sector organizations into the organizational networks necessary to carry out specific innovations, and creates market incentives for the use of new technologies.

Two points of central importance must be highlighted. Continuous synthetic innovation requires the sustained maintenance of organizational complexes that link a diverse set of public- and private-sector organizations. The United States has an inadequate complex supported by a special-interest policy system focused on winning in the manufacturing export market.

PART III

THE JAPANESE

ADAPTATION

THE synthetic reality had truly arrived when the Japanese developed a mature capability for continuous synthetic innovation of commercial products and processes that paralleled the United States's capability in defense, medicine, and agriculture. What the Japanese had by the 1980s was an ability to tap and gain synergism from a range of expertise. What that capability for synthetic innovation produced was a broad range of products and processes aimed at winning in the export race. What was initially the Japanese challenge is now one that includes a growing number of newly industrializing countries (NICs) intent on producing radical as well as incremental innovations.

CHAPTER 11

America's Commercial Competitors

$\mathbf{A}$ MERICA'S declining economic position is intimately tied to its loss of leadership in commercial high-technology products and processes. This occurred not because of an absolute loss of U.S. capability but because of the emergence of competitors with a superior capability. The Japanese capability to carry out continuous synthetic innovation crossed a threshold in the late 1970s.[1] In combination, the oil disruptions of the 1970s and America's bout with stagflation masked what was occurring.

Only during the latter half of the 1980s have we begun to appreciate that Japan represents the vanguard of nations that are developing capabilities for continuous synthetic innovation of commercial products. The challenge posed by the Japanese and the newly industrializing countries (NICs) can no longer be seen as a momentary problem resulting from a technological stumble by the United States. As Professor Daniel Okimoto, co-director of Stanford's Northeast Asia–U.S. Forum on International Policy, has written, "In such sophisticated technologies as very large scale integrated circuit (VLSI) memory chips and fiber optics, Japanese companies have succeeded in capturing large segments of world markets."[2] The Japanese are self-consciously seeking to carry out a range of radical innovations in such areas as biotechnology, electronics, and materials.[3]

Appreciation of the Japanese success requires understanding of how they evolved an organizational capability for continuous synthetic innovation aimed at the export market. The Japanese started with a significant advantage: they were not faced with the ideological strictures of the secular trinity. Perhaps on no point is there greater consensus than that, in Japan, the group takes precedence over the individual. Thus recognition that synthetic innovation requires groups and that group learning is of critical importance posed no ideological or cultural problem in Japan. On the contrary, this characteristic of the synthetic reality is perfectly in tune with Japanese culture.

Similarly, there is no ideological need to build walls of separation between the public and private sectors in Japan. Thus the notion of a free market—that is, a market free of or with minimal government intervention—has no deep roots in Japanese society. On the contrary, the Japanese believe that a free market sometimes leads to waste and inefficiency. Certainly the idea that the market will allocate resources to the appropriate economic sectors makes little sense to the Japanese. Rather, they assume that government has a key role in directing economic development. Chalmers Johnson, author of *MITI and the Japanese Miracle,* has characterized Japan as a developmental state and the United States as a regulatory state. He says Japan has a plan-rationale system, with government acting as the planner, while the United States has a market-rationale system, where individual buyers and sellers control events.[4] The public- and private-sector cooperation necessary for synthetic innovation is accepted practice in Japan.

Finally, although the Japanese have efficiency (low-cost production) as a primary goal, they appear to use a different set of time constants. Perhaps there is no clearer illustration of this general attitude than the slogan the Hitachi Corporation uses to communicate its purpose: "Though we cannot live one hundred years, we should be concerned about one thousand years hence." It is not uncommon when one visits Hitachi facilities to hear individuals ranging from workers in the plants to the company president refer to the slogan. This different perspective on time, and thus on the time period for measuring efficiency, gives the

Japanese a much different viewpoint from their U.S. counterparts. Perhaps the greatest difference is that the Japanese do not cut costs to achieve short-term efficiencies that then limit their ability to innovate and compete in the future.

In sum, the Japanese have a culture that views the group, public- and private-sector cooperation, and long-term innovation favorably. The key ingredients to success in the synthetic society fit comfortably with Japanese culture, while they run counter to the secular trinity in the United States.

Chalmers Johnson offers insight into the similarity of the elements of commercial success in both Japan and the United States. After noting the mechanisms of government/industry cooperation in Japan, he makes the following observation:

> This form of the government-business relationship is not peculiarly or uniquely Japanese; the Japanese have merely worked harder at perfecting it and have employed it in more sectors than other capitalist nations. The so-called military-industrial complex in the United States, to the extent that it identifies an economic relationship and is not merely a political epithet, refers to the same thing. If one were to extend the kinds of relationships that exist between the U.S. Department of Defense and such corporations as Boeing, Lockheed, North American Rockwell, and General Dynamics to other sectors of industry, and if one were also to give the government power to choose the strategic sectors and to decide when they were to be phased out, then one would have a close American approximation of the postwar Japanese system. The relationship between government and business in the American national defense industries—including the unusual management and ownership arrangements for the nuclear weapons laboratories and the existence of such official agencies as the former Atomic Energy Commission and the National Aeronautics and Space Administration—is thought by Americans to be exceptional, whereas it was the norm for Japan's leading industrial sectors during high-speed growth. It is also perhaps significant that aviation, space vehicles, and atomic energy are all sectors in which

the United States is preeminent, just as Japan is preeminent in steel production, shipbuilding, consumer electronics, rail transportation, synthetic fibers, watches, and cameras.[5]

Japan has, then, a cultural ideological environment that is more compatible with the synthetic society than is the case in the United States.

Organizational Forms

In the following section we pursue an understanding of how the Japanese developed their broad-based organizational capability for continuous synthetic innovation. This requires looking at the post–World War II development of Japanese manufacturing. Tracing the Japanese evolution from World War II devastation to world leadership in commercial innovation offers two important insights. The first is an appreciation of the Japanese challenge and the critical role that organizational capabilities play in that challenge. The Japanese ability to synthesize commercial technologies rests on the same capability to tap and integrate expertise, skills, and information that distinguishes the U.S. defense, medical, and agricultural complexes. Japanese organizational arrangements, however, are quite different from those in the United States. Second, the Japanese model of economic/technological development has become the explicit reference point for a number of NICs. If successful, the NICs will become increasingly serious competitors in the high-tech commercial marketplace. The evidence suggests the NICs have cleared the early hurdles in the evolution to a capability for continuous synthetic innovation.

Some cautionary observations are useful before looking at Japanese technological development. For most of the post–World War II period, the Japanese have played technological catch-up, and it is much easier to design policies and manage institutions when a successful model exists. It remains an open question whether the Japanese will be successful now that they are at the technological frontier and are thus without obvious technologies to adopt and improve.[6] Further, even with a clear picture of which product areas offered growth, the Japanese success did not follow a smooth, error-free path. Experimentation, luck, and the unique characteristics of Japanese culture (that freed it from the strictures of the secular trinity) may explain as much of the success as thoughtful, rational societal planning and management.

During the post–World War II period manufacturing for the international market has been Japan's dominant focus. Only the ability to compete internationally would enable Japan to pay for its essential imports.[7] Thus, as noted by Professor Okimoto, for the Japanese choices about what was manufactured could not be left to the free market.

> If the composition of the emerging economy is left entirely in the hands of the market, the government runs the risk that finite resources will be diverted from key industries like semi-conductors, computers, and telecommunications—on which Japan's future competitiveness hinges—and invested instead in sectors of less importance for the industrial economy (restaurants, coffee houses, and real estate speculation for example).[8]

This focus on industries capable of winning in the international marketplace has also characterized policy in the NICs.

Evolution: Industrial-Synthetic

The industrial-to-synthetic evolution of Japan has had four distinctive foci: (1) low-capital, labor-intensive goods, (2) heavy industry, (3) mass-produced, sophisticated consumer goods, and (4) radical synthetic innovation.

LOW-CAPITAL AND LABOR-INTENSIVE

Immediately following World War II, an industrially devastated Japan, under the close oversight of the Supreme Command, Allied

Powers (SCAP), focused its primary attention and limited capital on exporting low-capital, labor-intensive goods. Textiles and clothing were among Japan's first major export successes and also the first products to trigger political efforts in the United States to limit imports in the postwar period. SCAP played a key role in focusing Japan on export-oriented manufacturing. In the first years following surrender, Japan's Reconstruction Finance Bank subsidized investment in coal and electric power. Under direction from SCAP, that subsidy had been withdrawn by 1950, which led to a radical reallocation of the limited private capital. Chalmers Johnson has described what occurred as follows: "Funds for coal and electric power development declined drastically while funds for the reestablished textile industry shot up. SCAP was pleased by this development since textiles earned foreign exchange."[9]

What began with the focus on textiles and other low-capital, labor-intensive manufactured products in the late 1940s established a pattern that has continued to the present. Japanese manufacturing has consistently focused on those product areas seen as offering the greatest international competitive and comparative advantage. Over the postwar period, the type of manufacturing that could provide competitive advantage has changed with the changing mix of products involved in international trade, the growing strength of the Japanese capital position, the rise in labor costs, and the increasing educational and organizational capabilities within Japan. That nation began the evolution to a synthetic society with significant strengths, a relatively well-educated and highly motivated work force, and a social structure that allowed it to take advantage of the strengths of its population. Most of the NICs have also launched their industrialization by focusing on low-capital, labor-intensive manufactured goods—most frequently textiles, clothing, and shoes—and most have enjoyed relatively well-educated, highly motivated workers and social structures that have allowed their workers to be utilized effectively.

HEAVY INDUSTRY

The second focus was on capital-intensive heavy industry. Japanese competitive advantage in heavy industry came from mating state-of-

190

the-art production processes with a highly skilled, low-cost labor force that used the lowest-cost raw materials available on the world market.

Steel and shipbuilding represented high-priority cases of a deliberate national policy aimed at rapidly adopting state-of-the-art technology. The centrality of this policy is suggested by the fact that, as Professor Leonard Lynn has stated, "until the late 1960s much of the emphasis of Japanese research and technology policy was on finding and intro-ducing the best foreign technologies."[10] The Japanese government played a central role in guiding this search through the Ministry for International Trade and Industry (MITI). MITI assured the nation of low-cost access to state-of-the-art technologies by keeping individual Japanese companies from competing with each other for the right to use those technologies.[11]

Having acquired state-of-the-art production technology, the Japa-nese immediately began to carry out continuous incremental innova-tions aimed at delivering both higher quality and lower cost. Innovation was delivered by production people and organizations, not by separate R&D organizations. Professor Ken-ichi Imai of Hitotsubashi Univer-sity provides insight as follows:

> Japanese as individuals and as members of firms are very aware of the importance of technology, and there is a widespread grass roots basis supporting the spread and development of new and better ways of doing things. This is one part of the accumulation of small innovations for which Japanese firms are renowned.[12]

The ability to build into production organizations the search for continuous incremental-process innovations proved to be a powerful competitive force. As process innovations reduced costs and increased quality, Japanese products became more and more competitive in the world market. Initially, the Japanese steel in the international market-place was a basic product that was competitive because of price. Simi-larly, early Japanese ships were unsophisticated and low-cost. Over the course of the post–World War II period, however, the increasing sophistication of Japanese basic industries and continuous incremental innovation has resulted in Japan's moving to a position in the 1980s where it is among the world's largest producers of specialty steels and sophisticated ships.

As this evolution from bottom-of-the-line to top-of-the-line production capabilities evolved, the companies established and then expanded their research and development capabilities. Although R&D capabilities have grown more sophisticated, they have retained a primary focus on market needs and opportunities. Japanese R&D goals have included lower costs, higher quality, and superior or market-specific performance capabilities. R&D has aimed at responding to or creating international market opportunities. On the other hand, while the Japanese basic industries were adopting state-of-the-art production technologies and consistently improving them, many of America's basic industries were becoming increasingly obsolete in technological terms. American commercial shipbuilding has effectively disappeared, and during the early 1980s the lack of competitiveness of America's steel giants was documented in the media by bankruptcies, plant closings, and company diversification. Many factors go into explaining the relative competitiveness of basic industry in Japan vis-à-vis the United States, but none is more central than the greater willingness of the Japanese to adopt state-of-the-art technologies and then to improve them continuously. Again, the patterns established by the Japanese are being repeated in Korea, Taiwan, and other NICs.

SOPHISTICATED CONSUMER GOODS

The third focus in the evolution of Japan from an industrial to a synthetic society was on sophisticated products and technologies aimed at the mass markets of the world—for example, cameras, watches, pianos, and pickups. Consumer electronics is the classic example of Japan's movement from total lack of participation in the international marketplace to dominance in it. Here, as in the case of heavy industry, the initial Japanese approach was to license or buy the right to produce state-of-the-art products. A key event was the identification in the early 1950s by Sony (at that time Tokyo Tsushin Kogyo) of the potential market for products based on the transistor and, more broadly, on solid-state electronics.[13] Sony purchased from Western Electric the right to use the technology with the recognition that the transistor's performance would have to be significantly improved. This was a symbolically critical turning point. The need for incremental innovation was driven by the desire to produce a product (the pocket radio) so different that there were serious questions as to whether a market existed.[14] The incremental innovation of a transistor with superior performance became the basis for a growing menu of consumer electronics products. The Sony example became a model for a growing number of Japanese companies that were developing their own capabilities for carrying out continuous, incremental product innovations. Many of the products traced their lineage to commercial spinoffs of U.S. defense innovations. Through purchasing and licensing arrangements the Japanese plugged into the same defense synthetic system that had given American commercial producers their dominance during the first decades following World War II.

Japanese manufacturing organizations are dominated by engineers and technical workers. Perhaps for this reason the distinction between research and development and manufacturing and marketing is much less pronounced than in U.S. companies. Engineers dominate, but their goal is not technological elegance; it is competitive advantage in the international commercial marketplace. It is assumed in Japanese companies that engineers involved in product development will frequently lead marketing specialists in the identification of sales potential. For example, Akio Morita, President of Sony, has noted that had a market survey been taken, it would probably have recommended against the highly successful Walkman.[15] I have frequently asked Japanese businessmen if they prepare discounted cash-flow analyses—a given in most American companies—before introducing new products. Most Japanese indicate that they see little value in such analyses for new products since there are no reliable market numbers available.

As the Japanese developed manufacturing sophistication and the capacity to innovate both products and processes incrementally, their impact on the international market became something like a tidal wave. This process reached critical mass in the late 1970s. By then, many Japanese products consistently had three advantages: lower cost, higher quality, and superior performance. No great conceptual breakthroughs or inventions were behind the success. The source of Japanese competitive advantage was a pattern of steady evolution. Increasingly, Japan's competitive strength was the result of a superior organizational capacity for incremental innovation.

Incremental innovation allowed the Japanese to restructure international competition. The world market was increasingly being driven by innovation-to-obsolescence cycles. With each incremental innovation delivering one or some combination of lower cost, higher quality, and superior performance, the Japanese were making their competitors' products more and more obsolete.[16] As incremental innovation both created new markets and reduced foreign competition, the market for Japanese goods exploded and they began to enjoy the additional benefits of economies of scale. To meet the demand for products that developed rapidly and were quickly replaced by demand for the next innovation, it was necessary to have flexible manufacturing systems that could change from one product to another rapidly and at low cost.[17] By the 1970s Japan had companies capable of continuous commercial innovation in a large and growing number of product sectors with financial and marketing power equal to or greater than the largest American and European corporations.

Prior to World War II and in the years immediately following, Japan had a reputation for producing low-cost, bottom-of-the-line, shoddy goods. In the 1960s and 1970s, however, Japanese corporations were recasting their products' image. In combination, the wide use of the statistical quality-control techniques developed by Deming and others and a consistent pattern of higher-performance products saw Japanese products selling for more than American products.[18] By the 1970s, too, the Japanese were deliberately selling top-of-the-line products. The Sony Trinitron television was marketed at prices substantially higher than competing American-made televisions, yet American buyers were bragging about buying the Sonys. In the jargon of marketing, a growing number of Japanese products were "up-scale" because of their higher cost, higher performance, higher quality, and the higher status they conferred on buyers.

By the late 1970s, Japan had become a juggernaut for sophisticated consumer goods in the world market. Together, incremental product and process innovations allowed the Japanese to drive the innovation-to-obsolescence cycles for consumer products more and more rapidly. The instruments of competition had gone beyond price and style. By now the Japanese were playing the role in the commercial high-technology market played by the United States in the arms race. They were incrementally producing products with all or some combination of the qualitites of superior performance, superior quality, or lower cost ahead of their competitors, and they were doing it repeatedly, cycle after cycle.

During the same period the NICs, too, began to move from a primary emphasis on basic industries into sophisticated consumer products. Like the Japanese before them, they took advantage of li censes from other countries and used their skilled, low-cost labor forces to produce products aimed at the low-price end of the market. The NICs' developing focus on sophisticated consumer products provided additional incentive for Japanese companies to shorten the time involved in innovation-to-obsolescence cycles. As labor costs rose rapidly in Japan, their initial labor-cost advantage was shifting to the NICs. With Japanese labor costs continuing to rise, the NICs had a competitive advantage, and the Japanese had a powerful incentive to seek different advantages in their superior capabilities for innovation.

RADICAL INNOVATIONS

The 1980s have seen an increasing Japanese concentration on the fourth focus, radical innovations. Two things drive this focus: (1) growing evidence that the NICs can compete in the manufacture and incremental innovation of sophisticated high-tech products; and (2) recognition that the technological gap with the United States has been closed in most areas of commercial high-tech products. Continued growth and leadership can no longer rely on adoption of foreign inventions or innovations that are then incrementally improved.[19]

Perhaps the period of radical innovation in Japan can be symbolically pegged to the introduction of the videocassette recorder (VCR). Initially introduced to the mass market by Sony, a company with an overt policy of winning in the economic marketplace by being first with superior performance and quality products, the VCR triggered an explosion in demand. Lester Thurow, dean of MIT's Sloan School of Management, has summarized what happened.

In just one month, October, 1984, Japan made more than 2.6 million video recorders (VCRs). How many video recorders were being made in the United States at the same time? Not one. Some are marketed under American brand names but not one has been made in America—100 percent imported.[20]

The radical innovation represented by the VCR was not the result of Japan's developing the first prototype aimed at the mass market. As Abegglen and Stalk have pointed out in *Kaisha,* their powerful analysis of Japanese corporations, "At least two Western companies had prototypes in hand at an earlier date than their Japanese competitors."[21] The key to the success of the VCR, as is true for new products generally, was the ability to produce distinctively new designs and then achieve high-quality, large-scale production rapidly. In sum, Japanese success with radical innovations is largely the result of the same factors that produced successful incremental innovations—Japanese corporations consistently deliver products to the international market more rapidly than their Western competitors. A critical ingredient is the more effective communication and transfer of expertise within Japanese companies—the more effective ties among design, engineering, manufacturing, and marketing.[22]

In the same way that the defense complex in the United States has provided leadership in radical defense innovations, the Japanese are developing organizations with leadership in radical commercial innovations as one of their goals.

Japanese Organization

Contrary to popular mythology, the Japanese government does not play a role in commercial innovation similar to that of the U.S. government in defense. Furthermore, the role of the Japanese government in guiding and stimulating manufacturing has been declining steadily since the early 1970s.[23] MITI did play a very significant role in stimulating and guiding the development of basic industries, but its role in the development of the sophisticated consumer-products sector has been less important. The role of the Japanese government in the focus on radical innovations is also limited, although it is perhaps critical sometimes in the early identification of areas of importance.

Japanese corporations are the dominant organizations in commercial innovation. Their organizational evolution is the lead story of Japan's present competitiveness. Recall that the first major focus on innovation occurred in basic industries. During the early phase of development of Japanese heavy industry, the companies devoted very limited resources to R&D. In the beginning, innovation occurred as an integral part of production activities. Only in the 1960s did there begin to be a distinct focus on R&D.[24] With the focus on sophisticated consumer products, distinctive planned R&D began to grow. Initially, R&D was very applied. R&D personnel were primarily engineers and technicians who were focused on meeting market opportunities. As Japanese companies increased the range of products that were the object of innovation efforts and as they sought to push the state of the art greater distances, it was necessary to build a broader R&D base. Corporate R&D expanded to include applied research as well as development. The expected payoff time for R&D was extended, but the hope was that higher payoffs would flow from major advances. Some measure of what began to happen to R&D in Japan in the late 1960s is illustrated by table 11–1.[25]

Table 11–1

Real Increase in Research Expenditures (1975 = 100)

	Japan	United States	West Germany
1965	43	96	53
1970	87	102	83
1975	100	100	100
1980	130	125	132
1982	152	128	133

SOURCE: Kagaku Gijutsu Cho Hen (Science and Technology Agency), *Kagaku Gijutsu Hakusho* (Science and Technology White Paper on International Comparison and Future Themes), (Tokyo: Okurasho, 1982), 104. From James C. Abegglen and George Stalk, Jr., *Kaisha: The Japanese Corporation* (New York: Basic Books, 1985), 125.

Recall that complex technological innovation that pushes the state of the art a substantial distance normally requires the ability to tap and integrate skills ranging from fundamental research to marketing across a broad range of substantive areas (see chapter 3). Increasingly, Japanese companies have been expanding their in-house expertise both in the direction of fundamental research and across a broader range of substantive areas. The recent importance that Japanese companies (as compared to U.S. companies) have given to R&D capabilities is reflected in table 11-2.[26] Manufacturing companies contain the primary innovation capability in Japan. They have large and broad-based in-house R&D capabilities, plus the range of manufacturing and marketing expertise needed to carry out innovations. Industry provides roughly 69 percent of R&D expenditures in Japan;[27] the Japanese government provides the rest.

Innovation in Japan is primarily organized and managed by corporations, and they play the integrative role played by government in the United States. In part this reflects the fact that expertise in Japan is not fragmented among such a large number of organizations as it is in the United States. The concentration of a broad range of expertise within individual corporations also assures that R&D remains focused on the primary objective—winning in the international marketplace. In sum, the primary location of Japanese R&D within the corporations

Table 11-2

Research and Development as Percent of Sales (1983)

U.S. Company	R&D (%)	Japanese Company	R&D (%)	Difference
General Electric	3.4	Hitachi	7.9	+4.5
General Motors	3.5	Toyota	3.9	+0.4
Eastman Kodak	7.3	Fuji Photo Film	6.6	−0.7
DuPont	2.7	Toray Industries	3.1	+0.4
U.S. Steel	0.5	Nippon Steel	1.9	+1.4
Xerox	6.6	Canon	14.6	+8.0
Texas Instruments	6.6	NEC	13.0	+6.4
RCA	2.4	Matsushita Electric Industries	7.2	+4.8
Goodyear	2.6	Bridgestone	4.5	+1.9
Eli Lilly	9.7	Shionogi	9.6	−0.1

SOURCE: Adapted from "R&D Scoreboard: 1983," *Business Week*, 9 July 1984, 63–75; and *Nikkei Kaisha Joho*, no. 3 (Nikkei Company Information) (Tokyo: Nihon Keizai Shimbun, 1984). From James C. Abegglen and George Stalk, Jr., *Kaisha: The Japanese Corporation* (New York: Basic Books, 1985), 120.

that manufacture and sell in the international market has much the same focusing effect that the large DOD R&D expenditures and market purchases provide for defense in the United States.

Expertise Location

Following is a sketch of the location of expertise in Japan and a look at how Japanese companies use R&D to meet the challenges of the changing international market. Japanese R&D capability is primarily located in three organizational settings: universities, government laboratories, and industry.

Japanese universities do not play the powerful research role of American universities. Further, their research faculty are not tightly tied, through common substantive interests and funding, to mission-oriented government agencies and industry in the way of American university faculty. The vast majority of funding for research in Japanese universities is provided by the Ministry of Education, Science, and Culture. The Ministry controls half of all government R&D funds, and two-thirds of its R&D monies go to universities. The general purpose of the Ministry's support for universities is to ensure an adequate supply of research and technical manpower, rather than to fund specific substantive research.[28]

Japanese companies and government research laboratories employ university graduates with the expectation of providing the specialized training and opportunities necessary to make them researchers. As pointed out by economist Gary Saxonhouse of the University of Michigan, "While a Ph.D. is almost a prerequisite for active participation in a U.S. corporate R&D laboratory, such an advanced degree is much less common in otherwise comparable Japanese facilities."[29] The Japanese have substituted in-house research and training within the corporation for what in the United States is a university function. Doubtless, the fact that most research and training of researchers occurs in the corporate context is another factor in the strong R&D orientation on products and processes aimed at the international marketplace.

Most ministries have research laboratories. However, the Ministry of Education, Science, and Culture; the Science and Technology Agency; and the Ministry for International Trade and Industry contain the overwhelming portion of the government's R&D laboratory capability.[30] These laboratories exist to assist the ministries in carrying out their unique missions, but recently there has been a major effort to ensure collaboration and communication between the laboratories and industry. The goal appears to be to have government laboratories partially serve the function that universities serve in the United States.

The central role of Japanese industry in science and technology is reflected in the makeup of the Japanese Government Council for Science and Technology. The Council consists in part of the prime minister plus the heads of a number of the key science- and technology-related ministries or agencies. As underlined by Leonard Lynn,

> The other members are the Chairman of the Japan Science Council and five members of "outstanding ability" from the scientific community. These members of the "scientific community," it should be noted, also provide a linkage to big business. In 1983 two were chairmen of the boards of major firms and the third was the head of an industry association.[31]

The size of the research and development capability being built within Japanese companies is impressive. In 1965 Japanese R&D expenditures represented about 6 percent of U.S. R&D expenditures. During the 1983–86 period, the Japanese were spending 2.8 percent of their GNP on civilian-oriented R&D, the highest percentage of any country in the world.[32] With the world's third-largest GNP, Japanese R&D expenditures are presently large, and the national objective is to devote 3.5 percent of GNP to R&D.[33]

Although the cases of Sony, Honda, and Casio are sometimes used to illustrate the fact that part of Japanese postwar economic-technological success is the result of entrepreneurs, there are few recent instances of such entrepreneurs. Innovation in Japan occurs mostly in large, multiproduct companies like Sony, Toyota, and Hitachi—not by Silicon Valley–like small companies. In general, the large Japanese companies seek to be self-sufficient, either in-house or in their immediate supply and distribution network, so far as the range of expertise and skills needed to generate and then carry new ideas to the market is concerned.

In the past, however, Japanese industry has done little in areas of fundamental research where the market potential of the work has not been clearly established. As the Japanese have come to believe that their future economic well-being rests on the economic rewards of radical innovations, the country's basic research weakness has become a focus of growing concern.[34] In the absence of strong graduate research universities similar to those in the United States, the Japanese have sought to deal with this deficiency in a number of ways.

First, a variety of initiatives have been undertaken to stimulate research in the Japanese universities and to link the universities more closely with industry. Probably the most important of these initiatives has been the availability of funds for specific kinds of research; these funds are allocated, at least in part, on the American competitive-grant or contract model. This approach appears to be rather limited and is not viewed with much enthusiasm by many Japanese academic researchers. The Japanese have also sought to provide a model of a new kind of research university with the establishment of the University of Tsukuba, which is modeled in part on American universities. Most observers, however, see the changes in Japanese universities moving slowly, so that it is difficult to imagine Japanese universities being able to play the role of American universities any time soon.

Second, most of the large Japanese corporations have, during the 1980s, given substantial emphasis to building greater organizational R&D strength aimed at fundamental research. In interviews, Japanese corporate managers consistently indicate that this focus on what they frequently refer to as "advanced research" is one that must continue. Their view is that research is essential for the well-being of their companies, but they also commonly suggest that the companies have a responsibility both to Japan and to the world to push understanding and scientific/technical capability. While real movement toward fundamental research has occurred, it still seems clear that the companies will be limited in the amount of fundamental or advanced research they carry out.

Third, government research laboratories have increasingly assumed a larger role in the development of fundamental research. Government research laboratories appear to be attempting to play part of the role of the research university in the United States. They are moving in the direction of more fundamental or generic research and, at the same time, are placing major emphasis on cooperation and joint research with industry. The science city at Tsukuba, which has two universities and fifty-two research institutes, has also become a focus of industry/government laboratory R&D coordination.[35] Most large corporations maintain some liaison or coordinating presence at Tsukuba with the expectation of enjoying the benefits of those government laboratories.

Fourth, both Japanese government agencies and industry expend major and continuing efforts to stay in contact with and participate in the international scientific/technological community. The Japanese are avid consumers of research and technological developments throughout the world. Indeed, the frequency with which Japanese scientists and engineers visit research and development facilities throughout the world is the subject of almost continuing comment. Japanese participation in professional conferences is large, and the number of Japanese who present scholarly papers is increasing. The Japanese have also sought to link into some areas by funding research, particularly at U.S. universities, in areas where they have an interest. Further, the Japanese appear to be eager participants, where they are not excluded, as industrial sponsors of U.S. university research centers, such as the NSF-sponsored Engineering Research Centers. Clearly, a major purpose of all these efforts is to ensure that Japan will remain abreast of the research frontier and be able to utilize new discoveries, knowledge, and information wherever they may occur in the world.

Finally, as the perception of need for radical innovations has increased, the Japanese government—particularly through MITI—has organized a number of major research projects for the purpose of bringing, as Professor Okimoto has noted,

> leading companies to work together on common projects of high national priority—avoiding wasteful duplication, pooling finite resources, and achieving research economies of scale.[36]

America's Commercial Competitors

Although particular projects may involve universities or government laboratories, the primary research strength being combined is in companies. The record with regard to these projects varies substantially, as does the eagerness of companies to participate. A key point, however, is that these major organized research efforts tend to focus on areas where it is widely believed there are great potential economic payoffs, (e.g., biotechnology, high temperature superconductors, the fifth-generation computer) and where research is at the "precompetitive" stage. Professor Okimoto describes the process as follows:

> MITI chooses which projects to organize and subsidize very carefully, in close consultation with industry. To qualify for government assistance a project must meet four criteria (1) the proposed project must be of seminal importance to Japan's technological progress and future economic well-being; (2) the research must be of the pre-commercial variety, so that participating companies do not gain decisive advantage over excluded firms; (3) government assistance must be indispensable for the project to get underway and be completed; (4) the time frame for the project's completion must be realistic. Government financing supplies the critical missing ingredient for launching projects of high capital costs and risks, relatively long gestation, fundamental technological importance, and broad commercial applicability.[37]

In carefully selected areas where precompetitive research is needed, the Japanese government plays the kind of linking role for companies that the Department of Defense has played in some cases in the United States. This government role has developed as the focus on radical innovations has become increasingly important. These major research efforts normally aim at the development of a body of knowledge that will make it possible to conceptualize and design commercially viable products or systems. Once the project has arrived at a point where a consensus exists on what commercial direction to pursue, the Japanese government withdraws, and the individual companies carry the effort to the market. The VLSI project is everyone's example of a successful effort of this type.

The importance and role of research and development for Japanese companies differs depending upon the company environment. Recent analysis suggests the following picture: those companies and industries involved in product lines that are experiencing rapid incremental innovation and rates of growth (growth industries), such as the electronics industry, tend to focus their primary emphasis on relatively specific product lines; thus R&D is aimed at incremental product innovation. This R&D focus reflects the fact that success comes from pushing the state of the art ahead of one's competitors or at least in parallel with one's competitors. Alternatively, companies operating in sectors that are experiencing little innovation and relatively low growth (declining industries) tend to focus R&D on diversification. That is, they use research and development and innovation to move into other or new product lines.[38]

Two things are distinctive about Japanese companies. Many are involved in a mix of businesses similar to U.S. conglomerates. In Japan, however, companies grow their own conglomerate capabilities. For a variety of reasons, it is not easy in Japan to expand into new product or market areas through mergers and takeovers.[39] Even if that were possible, the lack of entrepreneurial companies in Japan limits this route to diversification. Innovation is seen as the route to diversification for companies operating in stagnant markets. This has organizational consequences, and the first step in this process is to build into the company the capability for new or enhanced invention and innovation. The starting point is R&D capability. Professor Okimoto notes that organizational arrangements may vary.

> Japan's political economy is undergoing changes as the country's industrial structure shifts from the old-line industries to high technology. The nature and role of such institutions as the Keiretsu Groups, subsidiaries and subcontracting networks, and banking-industry relations are changing under the impact of such external forces as financial deregulation and fast-moving technology. A number of these changes are adaptations to the different functional requisites of high technology. The proliferation of small subsidiaries and the increasing shift of R&D from parent firm to subsidiaries, for example, ought to be understood as structural

adaptations to the requisites of innovation and compensation for the absence of a venture capital market.[40]

The key point is that innovation is seen both as a way to maintain leadership in rapidly moving areas and as a way to rescue declining industries—that is, a way to achieve diversification. Innovation requires building organizations that have research, development, production, and marketing capabilities. In Japan these capabilities tend to be built within fairly tightly knit organizational networks that are predominantly organized by major manufacturing companies or commercial-financial groups. The role of government laboratories and universities is focused primarily on providing fundamental or precompetitive research, and they appear to play only a limited role.

The Future Challenge

In the immediate post–World War II period, the Japanese began with a primary national goal of being competitive in the international marketplace because exports were essential to pay for the imports needed by a resource-poor country. In pursuit of that goal, the Japanese developed a capacity for continuous innovation through a process of evolution. Initially the focus was on a capacity for incrementally innovating process improvements and product improvements; then, by the 1980s, the Japanese were focusing major attention on carrying out radical innovations. The emphasis on radical innovations is not a substitute for incremental product and process innovation, however; rather, it is an addition. In the 1980s, then, Japan has defined its well-being in a way that has many similarities to the U.S. definition of the role of the defense technologies.

With the development of a broad-based and mature capability for continuous synthetic innovation of commercial products in Japan, the fundamental nature of the international marketplace has changed. Japan is now the major engine driving a sequence of commercial innovation-to-obsolescence cycles, and the speed and agility with which

Japan's companies have been able to drive these cycles has played a major role in making U.S. products uncompetitive. In sum, the Japanese have posed for the United States in the commercial sector the same kind of challenge that we have continuously posed for the Soviets in the military sector. The fundamental difference is that the Soviets have mobilized (so far with limited success) to meet the innovation challenge in defense, but the United States has not mobilized to meet the Japanese commercial challenge.

Some comparisons between the development of the synthetic capability in the United States and in Japan provide insights. First, in the case of defense innovation, the establishment of an organizational capability occurred in a manner analogous to the Big Bang theory of the creation of the universe. The development of the atomic bomb and the organizational system built under the Manhattan Project created an environment and a set of expectations that still permeate the American attitude toward innovation. All too frequently, U.S. innovation is approached as a big idea, a crash-program phenomenon.[41] Second, American innovation has involved a much larger financial role on the part of government than is the case in Japan. The government's role in the United States has been threefold: (1) it has involved a substantial portion of the front-end, high-risk R&D funding; (2) it has involved government's acting as the builder of the complexes and networks necessary to produce prototypes and to take those prototypes into production; and (3) it has involved the government playing a key role in market management, essentially creating market conditions either through purchases or by managing supply or by ensuring use, which has created the "pull" side of the continuous innovation process in defense, medicine, and agriculture. Finally, in the cases of defense and medicine, the primary focus of innovation has been on product performance. Cost has been a low-priority focus.

In Japan, on the other hand, the development of a capability for continuous synthetic innovation rose almost imperceptibly out of the nation's need to win in the commercial marketplace. If the Big Bang approach was the launching mechanism for U.S. innovation, in Japan innovation grew through an incremental and evolutionary development. The Japanese began with an emphasis on cost and then quality as the goals to be delivered by innovation. Their approach was to achieve those two goals and later superior performance through an endless series of incremental innovations. These innovations were, from the first, heavily focused on the international market. The Japanese approach to the market assumed that, with products resulting from innovation, they were in the process of creating markets, not just responding to them.

The role of the Japanese government in the innovation process has been much more limited than that of the U.S. government. The primary organizational locus of the innovation capability has been in large Japanese companies, and those companies have individually built their own innovation complexes and networks. The Japanese government does plays two important roles, however. First, it seeks to deal with the inadequacy of fundamental research capability in Japan. Second, it has created a broader environment that makes it possible for Japanese companies to take the long-range view needed to be competitive in an international marketplace that is structured around innovation-to-obsolescence cycles. Specifically, Japanese policy has made it possible for Japanese companies to be relatively unconcerned about short-term profits, and Japanese policy has assured large amounts of low-cost capital and highly qualified workers.

Two points deserve emphasis. The real challenge to Japan is that the NICs represent a large, growing, and ever more sophisticated capability for carrying out incremental process and product innovations.[42] Further, they have the goal of developing a capability for carrying out radical innovations that are focused on winning in the international commercial marketplace. For the United States the Japanese challenge has not plateaued. Indeed, the deliberate effort by Japan to accelerate its capability for radical innovation will pose ever more difficult competition for the United States.

PART IV

RESPONSE TO THE

SYNTHETIC REALITY

THE FUTURE well-being of the United States requires changes in both Americans' perceptions of how reality works and in public policy. Fundamental change without political/economic disruption is, at best, a hope, but the alternative is a near-certain American decline into second-class economic status. The following chapter sketches a set of basic actions that are necessary for America's future well-being.

CHAPTER 12

Needed Steps

THE UNITED STATES invented the organizational arrangements that have produced continuous synthetic innovation. This invention became the engine that converted an industrial society into a synthetic society. There is a powerful analogy between what happened with the videocassette recorder and what has happened with our broad capability to carry out continuous synthetic innovation aimed at the export market—the inventor has not enjoyed the economic benefits of the invention.

Japan and a growing list of other countries have developed superior capabilities for commercial synthetic innovation. While the United States remains preeminent in the innovation of defense, medical, and agricultural technologies, it no longer leads in commercial technologies, a fact reflected in our high-tech trade balance (see chapter 1). The fundamental change in the nation's trade position—and especially its *high-tech* trade position—has many causes. These include Americans' propensity to consume more than they produce; low savings rates; a deteriorating educational system; questionable management and labor practices; and our focus on the short term. At its core, however, is an irreducible fact—the United States has not built and sustained an organizational complex capable of continuous synthetic innovation aimed at the export race. We have not done in the commercial export sector what we have done in defense, medicine, and agriculture.

The arrival of the synthetic society came with the establishment in Japan of a broad-based capability for continuous synthetic innovation of export-oriented products and processes. With that development and the pursuit of similar capabilities in other countries, the nature of international commercial competition changed. Like the arms race, the export race became increasingly structured around innovation-to-obsolescence cycles. The number of products experiencing those cycles, and their rate of turnover, will likely increase as a growing number of NICs develop more sophisticated capabilities for synthetic innovation.

If the United States is to become competitive in the export race, it must build and sustain an organizational complex capable of synthesizing superior products and processes. Failure to develop such a capability will mean that the nation's ability to meet its commitments and satisfy its appetites will continue to erode. To date, Americans have been able to escape the consequences of their lack of competitiveness by heavy foreign borrowing. The pattern of borrowing already in place nearly assures a foreign debt of over $1 trillion by the early 1990s. It also appears a near certainty that at some point foreigners will refuse to loan money to feed U.S. consumption.[1] Thus if the U.S. is to escape a severe socioeconomic disruption, it must address its trade deficit.

Where is improvement possible? In 1987, total world trade amounted to about $3.0 trillion, with goods representing 83 percent and services 17 percent.[2] In contrast to the U.S. economy, which has experienced a steady growth in the portion of GNP represented by services, the split between goods and services in world trade has remained relatively constant. The export market has been and likely will continue to be goods-oriented. Seventy-five percent of U.S. merchandise trade is represented by manufactured products, with the export portion in 1987 consisting of 79 percent manufactures, 12 percent agricultural products, and 9 percent miscellaneous.[3] Given the world food glut and the tendency of most governments to protect their agriculture, the prospects for export growth in agriculture are limited. The conclusion seems compelling that if the United States is to achieve a surplus trade balance, it must be carried primarily by manufactures.[4] A more daunting challenge is hard to imagine. Former Secretary of Commerce Peter Peterson has described the challenge in chilling fashion.

Needed Steps

It is particularly inevitable that our net debt will reach the $1 trillion mark by the early 1990s no matter how vigorously we act to stem the inflow of foreign savings. Obviously, there are limits to the speed with which the United States can curtail consumption and generate growth in net exports. Consider, for instance, a scenario in which the United States, starting next year, makes steady additions to the value of its net exports such that its current account reaches zero by 1994 and its net debt is reduced to today's level by the year 2000. That sounds like a rather modest achievement. Yet it will still lead to a net debt of about $1 trillion by 1994 and will require a real improvement in the U.S. net exports of more than $20 billion a year, each year, for the next ten years, or a total positive shift of more than $200 billion. As Fred Bergsten has observed, the magnitude of the necessary adjustment facing us is equivalent to about two-thirds of our entire defense budget and is several times larger than the total shift resulting in the United States from the 1970s oil shocks.

According to the adjustment scenario above, we need to reduce our foreign borrowing stream by $20 billion yearly, or $200 yearly for each of our 100 million workers. Yet real net product per worker has been growing each year by just $135. Further, our continuing debt growth will mean that about $40 per worker per year must be devoted to rising foreign debt service payments. . . . The challenge facing America—generating a $275 billion positive swing in manufactured exports over the next decade—sounds tough enough without worrying about whether our trading partners will accommodate our necessities. Yet we often forget that our objective of huge yearly increases in U.S. net exports translates directly into decreases in the net exports of our major trading partners (recently the very source of much of their growth).[5]

If there is a possibility of making it through to the next century without a severe drop in our living standard and our support for defense and medical care, it must come from synthetic innovations aimed at the export market. If we are to escape a major economic disruption, we must out-innovate our commercial competitors.

A growing number of Americans believe that we must make major changes in our socioeconomic system. Proposed changes include: decreased consumption, probably achieved through consumption taxes; a balanced federal budget, achieved by cuts in spending and by increased taxes; increased savings rates and investments; decreased emphasis on short-term profits and increased emphasis on long-term competitiveness (implying restructured capital markets); more commercially oriented research and development; vastly improved education at every level; and greater job stability. All these changes are valuable and perhaps essential if the United States is to be competitive in the future. It is this book's central thesis, however, that even if all these actions are taken, they will still not meet the nation's challenge in the export race. *The central and irreducible requirement if the United States is to be competitive in the synthetic reality is that it must build an organizational complex with the capability of continuously producing commercially oriented synthetic innovations.* The nation must have an export race–oriented capability for synthetic innovation that parallels what it has now in defense, medicine, and agriculture.

Special-Interest Policy System

An organizational complex capable of providing continuous commercial innovation will require building a new special-interest policy system for trade and manufacturing. Thus the president and Congress must address the four generic requirements (discussed in chapter 5) for establishing a new special-interest policy system. They must:

(1)   establish clear national goals for trade and manufacturing;

(2)   establish the set of substantive activities (the sector) that will be the priority focus of trade and manufacturing policy;

(3)   establish the instruments that the special-interest policy community will have at its disposal for managing the substantive activities;

(4)   establish who will be the continuing participants in the trade and manufacturing special-interest policy community.

Needed Steps

Clearly stated goals are central to a policy for trade and manufacturing. The nation's manufacturing sector must develop a major focus on the export market. *Specifically, the president and Congress need to establish that it is a major goal of national policy to achieve a trade surplus in manufactured goods within a prescribed time period. Further, the goals should specify that primary emphasis be given to high-tech products and to innovative technologies and processes that can enhance the quality and lower the cost of low-tech manufactured goods.*

These goals reflect two compelling facts about the synthetic society. First, the U.S. economy is no longer separable from the world economy. It must be clearly recognized that competition in the world economy is national competition. Continuous synthetic innovation of commercial products and processes is critical to America's well-being. The other side of this reality is that the costs flowing from trade barriers (i.e., attempting to reestablish a self-contained economy) would be high. Therefore, trade barriers are not the answer. The answer is to produce superior products faster than our competitors. This means that the United States must drive the innovation-to-obsolescence cycles for manufactured goods in the same way it drives those for defense, medicine, and agriculture. Second, the need to emphasize high-tech innovation as a national goal reflects both the growing portion of manufactured exports represented by high-tech goods and the growing importance of high-tech products (e.g., robots) and innovative processes (e.g., just-in-time inventory systems) being used in the manufacture of low-tech products.

SECTORAL FOCUS

Implicit in the goals definition are the boundaries of the trade- and manufacturing-policy sector. *Specifically, the focus of the special-interest policy system should be on research, development, and manufacturing that produces "high-value-added" goods and services.* Two points deserve emphasis: First, major attention needs to be devoted to accelerating the speed with which both radical and incremental innovations are moved from concept through design and development to

215

production and marketing. Part of this process requires that significant emphasis be given to designing to inviolate cost standards. Roland Schmitt, president of Rensselaer Polytechnic Institute, highlights what is needed with regard to cost by comparing the Japanese approach to design with that of the United States:

> In the U.S. the design phase of the cycle of innovation concentrates on the features and performance of the product much more heavily than on the processes by which it will be manufactured. . . . A complex product, with many features and elements of performance, designed without regard to the intricacies of making it, becomes a product of high cost, questionable quality and uncertain reliability, however elegant and attractive it may seem in concept. Yet there is a strong propensity in U.S. engineering to follow this course. . . . By contrast . . . the Japanese are oriented to simplicity in their designs. . . . The Japanese designer designs to cost; however many design iterations there may be, the initial cost barrier remains inviolate.[6]

Doubtless the traditional U.S. approach has its origins in defense innovation, where performance was the dominant goal. Faced with the export race, public policy must focus major emphasis on simplicity and cost as well as performance; it must seek to modify the traditional focus of American design.

Second, public policy needs to emphasize process innovation as well as product innovation. Obviously a change in the focus of design is a factor, but the process focus must go further. Process innovation is a major route to both improved quality and lower cost. Concern with process innovation has been a growing concern of U.S. industry in the 1980s. All too frequently, however, this has led American companies to substitute machines (e.g., robots) for people. As Roland Schmitt has observed, the evidence seems compelling that what is needed "is not just the substitution of machines for people but a reconfiguration of manufacturing operations to most effectively utilize the combination of human effort and state-of-the-art automation technology."[7] This need to focus on process innovation, then, requires changes in the culture of manufacturing. The goal should be to create worker/man-

agement teamwork aimed at making those multiple incremental innovations that are central to improved quality and lower cost. Like so
much else in the synthetic society, a primary concern of process innovation must be time. A growing body of experience suggests that reducing
production time delivers both cost and quality benefits. When the time
required for what *Kaisha* authors Abegglen and Stalk call "work-in-
process turns" (i.e., the time required to carry out a function) can be
reduced, the other benefits follow.[8] Process innovations tend to be
driven by making reduced production time a priority and a continuous
objective.

POLICY INSTRUMENTS

The ability of the special-interest policy system for trade and manufacturing to build and sustain an organizational complex capable of
delivering continuous synthetic innovation requires appropriate policy
instruments. *Specifically, the policy system needs the ability to: (1)
provide up-front R&D funding; (2) link the diverse organizations (universities, nonprofits, government laboratories, FFRDCs, government
agencies, and industry) needed to carry out innovation; and (3) provide
incentives for marketing the innovations.*

The primary innovation problem the United States faces is not in
research and new ideas; it is in rapid development (prototyping), rapid
production, and rapid marketing. Several factors are evident here. An
obvious one is that the long lead times from concept to production that
characterize the United States are unacceptable given the pattern of
accelerating innovation-to-obsolescence cycles. In the present context—except in defense, medicine, and agriculture—government funding stops with research. Industry, faced with a need to deliver short-
term profits, is generally hesitant to risk large sums on longer-term
development and on the building of large-capacity manufacturing
capabilities in advance of predictable demand. Yet the failure to have
large production capacity in a market structured around innovation-to-
obsolescence cycles often leaves U.S. companies unable to meet a
demand that grows and declines rapidly.

Short of a fundamental restructuring of the U.S. financial system,
only a capability by government to assume major portions of the risk

associated with commercial development and manufacturing capacity will provide for U.S. competitiveness in the export race. In truth, the U.S. government is unique in its failure to underwrite the risks of commercial innovation. Frank Press, president of the National Academy of Sciences, recently spoke to one part of this increasingly recognized condition as follows:

> Regarding industrial research, just about every advanced country that we compete with provides some incentives for industrial R&D. We can say that's not our culture, but we're operating in a world where that is the culture in other countries. . . . It's tough to compete when France, Japan, and Germany do that.[9]

The need to link together the six organizational types with R&D, production, and marketing expertise is central to successful export-oriented innovation. This is a widely overlooked problem. Expertise in the United States is fragmented among a very large number of public- and private-sector organizations. Only the ability to rapidly construct organizational networks with the required expertise to innovate specific products will make the United States competitive in the high-tech market. Thus government must be able to do two things: (1) build and sustain an organizational complex and (2) facilitate the formulation of innovation-specific networks. At present, commercially oriented governmental involvement is politically unacceptable. But the problem is, in fact, even more severe. Present arrangements limit the ability of industry to tap expertise in government organizations. The difficulty in tapping and utilizing the huge body of expertise and skills located in government agencies, captive nonprofits, FFRDCs, and government laboratories is a costly limitation on commercial innovation.

As the complexity and inevitably the cost of development steadily increases, a major national capability is being bypassed. For example, many of the national laboratories have capabilities for providing prototyping that can only be matched by the largest of companies. An example is the Oak Ridge National Laboratory

> proposal for a prototype advanced microchip process facility. If, as some observers suggest, the case of advanced microchip process-

ing represents an example of the wave of the future, the capabilities of the Lab may become increasingly rare and increasingly important to competitiveness.[10]

Yet the barriers to the use of government capabilities for commercial innovation are presently so great that the capabilities are of very limited utility.[11] This remains the case in spite of recent legislation and supportive rhetoric.[12] Policy instruments must be provided that make government-owned or -dominated capabilities an integral part of the export-oriented innovation competition.

Finally, policy instruments that allow government to assist in the market competition must be designed. The international character of the commercial market precludes government's using the same instruments to create market incentives for innovations used in defense, medicine, and agriculture. Government can, however, do such things as assist in underwriting the risks associated with product and process innovation; provide long-term, low-interest funding for manufacturing capacity; help in the identification and development of markets; provide loan money for the foreign purchase of U.S. products; and make trade a central concern of foreign policy.

POLICY COMMUNITY

The last component necessary for the development of a special-interest policy system for trade and manufacturing involves structuring the policy community. *Specifically, the major participants at the federal, state, private-sector, and nonprofit research levels need to be identified or created and integrated into the policy community.*

At the federal level three organizational needs must be addressed. First, executive responsibility for trade and manufacturing should be consolidated in a new cabinet-level department. Second, congressional authorization and oversight responsibility should be consolidated in a single committee in each House. Third, a quasi-governmental National Corporation for Technological Innovation and Economic Competitiveness (hereafter referred to as the Corporation) should be created.

The creation of an effective, viable special-interest policy system for trade and manufacturing requires concentrating the presently fragmented policy responsibilities in both the executive and legislative branches. Only such concentration will make this policy area an equal of defense, medicine, and agriculture. In the absence of such a policy system, every new issue, problem, and technology creates a competition among existing special-interest policy systems for jurisdiction over those activities. The present situation makes international trade and manufacturing a sector over which the Departments of Commerce, Treasury, State, Defense—in truth, a seemingly limitless number of agencies and departments—struggle for jurisdiction under the "competition for growth rule." The pattern is duplicated by congressional committees (see chapter 5). In this situation, policy is inherently unstable and slow-moving. Policy instability precludes the development of the essential ingredient for competitiveness in the export race: an organizational complex capable of carrying out continuous synthetic innovation. Such organizational complexes require powerful, stable, supportive special-interest policy systems.

The Corporation, the third element of the federal organizational needs that must be addressed by the president and Congress, should have the specific mission of building and sustaining the needed synthetic-organizational complex. Specifically, it should provide the primary government support for commercial R&D; act to link public- and private-sector organizations into a complex and (where needed) into networks; and, where possible, it should assist in creating conditions that facilitate the provision of manufacturing and marketing capabilities.

The Corporation must have the widest possible latitude in using its financial resources to promote commercial innovation. It should be able both to initiate proposals and to respond to proposals from industry, universities, and the various nonprofit and government R&D organizations. It must be capable of acting like the World War II Office of Scientific Research and Development. Moreover, the Corporation should be free to fund activities ranging from R&D to long-term loans for investment in production facilities and for purchase of innovations. Furthermore, it is essential that the Corporation be free to use procedures and funding techniques of its own design; it must be free of the existing burdensome requirements of government contracting and funding. In sum, it must have the capability to move rapidly, flexibly, and with stability and predictability.

To have these characteristics, the Corporation should be governed by a board of directors appointed by the president for twelve-year terms staggered so that one board member is appointed each year.

Finally—and this is essential—the Corporation should have a funding base independent of annual congressional appropriations. A secure, stable, large funding base is necessary to assure the Corporation of the resources and long-term stability necessary to build the required capability for continuous commercial innovation.

Given the belief of Americans in the secular trinity, many will see the proposed Corporation as un-American heresy. Nevertheless, the synthetic reality requires that the United States build a broad-based capability for commercial innovation. Experience suggests that only federal government leadership and support can make that possible.

The key participants in the trade- and manufacturing-policy community at the state level are of two generic types. One is the traditional state departments of commerce. The other is the newly created state agencies or units that have technologically based economic development as their mission.[13] Nearly every state has created such organizations in recent years. They provide a potentially important set of participants in the special-interest policy system for trade and manufacturing, and they also represent potentially important participants in the organizational complex for innovation. In a recent book Schmandt argues that these state initiatives may represent a major new source of both policy- and technology-based economic creativity and initiative.[14] When joined into a policy system and a national innovation complex, they may be particularly effective in facilitating high-tech competitiveness.

The range of participants in the private sector is large and includes the critical manufacturing/marketing segment of the innovation complex. The recent economic shocks experienced by many companies have created an awareness of the need for innovation as well as a growing focus on the international market. Thus the private sector may be increasingly sympathetic to participation in a special-interest policy system for trade and manufacturing.

Finally, the nonprofit R&D organizations, especially the universities, are already focusing growing attention on trade and manufacturing.

The need to build in the universities a strong and continuing R&D base that is focused on commercial innovation and the export market seems evident. The universities need to be linked to and provide input on trade and manufacturing the way they do in defense, medicine, and agriculture. In the American system, universities are not only an important source of ideas for innovative technologies, they are also a source of ideas and analyses important to policy.

Unanticipated Consequences

Although the creation of a special-interest policy system for trade and manufacturing and an organizational complex capable of continuous synthetic innovation are critical to competing in the export race, another issue must also be addressed: some synthetic innovations generate negative unanticipated consequences (see chapter 1). The recognition that technologies can create negative unanticipated consequences poses three serious challenges: How do we inform policy making of unanticipated consequences? How do we deal with negative unanticipated consequences? How do we deal with the expectation of unanticipated consequences?

Until the 1970s, decision making concerned with technological innovation in both the public and private sectors was made with little awareness of potential unanticipated consequences. This was because there was little public awareness of a link between innovation and negative unanticipated consequences. By the 1970s a growing body of research had demonstrated cause-and-effect links, and concern with those linkages became a common component of policy making. During the 1970s and 1980s, however, the costs of unanticipated consequences became so evident that major efforts were devoted to their identification. Government agencies, as a result of the National Environmental Policy Act's requirement for Environmental Impact Statements, became much better informed about unanticipated consequences.[15] At the same time, corporations were exerting much greater efforts to inform themselves of unanticipated consequences because of potential financial liability. In combination, the awareness of likely unanticipated consequences and increased professional capabilities to identify them resulted in information about unanticipated consequences becoming a standard input into decision making.

Over the last two decades the policy system has developed reasonable mechanisms for addressing unanticipated consequences. Private-sector organizations that can be identified as the cause of negative impacts can be held financially liable by the courts and regulatory agencies. Where responsibility cannot be identified, where the responsible party has inadequate resources, or where government is responsible, mitigation has become a government responsibility. The classic illustration is the Superfund, established to clean up toxic waste sites. Although the slow processes of government are frequently criticized by those who see themselves as damaged, it is not evident that a better process can be formulated.

On the other hand, the problems associated with the by now general expectation of unanticipated consequences have not been effectively addressed and call for public-policy action. Here the problem is the constraining effect the expectation of unanticipated consequences has on innovation. The major problem is with those technologies aimed at the commercial marketplace. The potential liability associated with negative consequences has had a dampening effect on innovation. If the key to winning in the synthetic society is getting there first or early, anything that increases the time from conception to marketing is damaging. Surely the effect of events like the Dalkon Shield experience must give all innovators pause.

Dealing with the dampening effect that potential liability from unanticipated consequences has on innovation will require legislation that provides for the following: *a maximum liability should be established for damages resulting from negative consequences of new products and processes.* The model should be the Price-Anderson Act, passed initially in 1957, which established a liability ceiling for nuclear power–plant accidents.[16] *The needed legislation should provide a government liability insurance program to cover any difference between the liability ceiling and the coverage limit provided by private insurers.*

To obtain the protection of this liability system, a company would need to do the following: (1) contract with an approved independent

assessment organization for a study of potential unanticipated conse-
quences of the innovation; and (2) submit an application for liability
coverage and a copy of the study to an expert panel. Panel members
should be economically disinterested and represent the range of expertise
necessary to judge possible unanticipated consequences. In addition,
the review panels should be mandated to deliver a decision on whether
the liability protection will be provided within three months of the
application. Failure to act should result in automatic liability coverage.
Finally, it will be necessary to establish a new organization to handle
the convening of the panels. This should be done with the advice of
the National Academies of Science and Engineering. The academies
have great experience in tapping the nation's expertise for assembling
panels to review specific scientific/technical questions.

The purposes of the proposal for liability protection are twofold. One
is to ensure that an informed and prudent judgment is made concerning the societal risk that may result from unanticipated consequences.
The other is to provide a predictable level of risk for companies involved in synthetic innovation. There should be no requirement that
innovators have liability protection; the choice of whether to seek
protection should rest with them.

Resource Reallocation

If the United States is to develop a broad-based capability for continuous synthetic innovation aimed at the export market, large resources
will be required. Obtaining the needed innovative resources will require
reallocating some of the resources presently devoted to defense, medicine, and agriculture. The largest reallocation will have to come from
defense. Recall that two powerful and opposing forces have the United
States in a vice: in defense, medicine, and agriculture the organizational
complexes are continuously creating an expanding menu of complex,
high-cost technologies; and, at the same time, the rapidly growing
capacity for commercial innovation in other countries has left the
United States with a large and growing trade deficit. To date, the

nation's response to this squeeze has been twofold: to gain relief through foreign borrowing (a short-term response with disastrous long-term consequences) and to seek answers in greater efficiency in defense, medicine, and agriculture (a response that is not working).

In the case of defense, the efficiency techniques have been to change contract terms, increase competitive bidding, and provide for more detailed management. When these responses have not provided solutions (i.e., lower costs), the pattern has been to stretch out development times and buy smaller quantities of the weapons systems in the production phase. The overall effect appears to be both a slowdown in innovation and an increase in the unit costs of weapons systems.

In medicine the same essential responses have been occurring. Increasingly the effort has been to make health care a private-sector activity, with the expectation that competition will lead to efficiency. Government has sought simultaneously to provide for cost control through tighter management and more and more detailed regulations. Nonetheless, medical costs have continued their upward spiral.

In the case of agriculture, there has been a costly stalemate. In part, the market has been allowed to operate and drive out inefficient farmers through bankruptcy, while at the same time federal subsidies have increased by an order of magnitude, from $2.8 billion in 1980 to $23.1 billion in 1987.

Within the special-interest policy systems for defense, medicine, and agriculture the pattern of inertial innovation has created increasing demand for resources. Decision making in special-interest policy systems is structured to protect existing resources and distribute new ones. There is an inherent inability to select innovations that benefit some participants in special-interest policy communities while rejecting innovations that benefit others. Special-interest policy systems are incapable of carrying out major resource reallocations. Yet if the United States is to become competitive in the export race, it must devote much larger resources to commercial innovation. Reallocation, both within and among special-interest policy systems, can only occur if decisions are made in the presidential/congressional arena. The critical resource-reallocation decisions are those concerned with the development of major technologies.

Defense is the sector that is of primary concern because of its size and its growing number of technological options and their rising economic and scientific/technological costs. For example, the present system has deployed seven distinctive strategic delivery vehicles: (1) bombers, (2) ICBMs, (3) IRBMs, (4) SLBMs, (5) land-based cruise missiles, (6) ship-based cruise missiles, and (7) airplane-launched cruise missiles. The first four types have gone through several generations, and experience suggests that a similar pattern of incremental innovation will follow for cruise missiles. Thus it is likely that new types of delivery vehicles will be developed in the future.

In an environment of shrinking resources, the likely future will be one in which smaller numbers of each incremental innovation will be deployed in order to stay within budget limits. Because of a loss of the economies of scale, unit costs will be higher. At the same time, innovation will be stretched out to keep annual development costs within the bounds of annual budget availability. The result will be higher development costs, but, more important, longer innovation-to-obsolescence cycles. Stated succinctly, the critical cost will be in time, the variable that is most important to success in the synthetic reality. In the immediate future both the nation's defense capability and its commercial competitiveness require decisions that will result in deploying a larger number of a smaller set of weapons systems. Fundamentally different allocations decisions must be made, and they can only be made by the president and Congress. Given the nature of the American policy-making system, it is not easy to design mechanisms that will force the reallocation decisions required by the president and Congress. In the end, the forcing mechanisms will be the rising foreign debt and the refusal of foreigners to continue loaning money to the United States. Lester Thurow suggests what will likely happen.

> Eventually lenders quit lending, and when this happens painful changes in lifestyle are forced on the borrower. This happened in Mexico in 1982 and it will happen to the United States at some point in the future."[17]

At this point there will exist an opportunity for addressing the need for resource reallocation. The choice will be difficult, but it will be critical to the nation's future. The ability to compete in both the arms race and the export race will rest on difficult presidential/congressional decisions concerning the division of the nation's resources between commercial and defense innovation. Those decisions will be structured around choices involving major weapons systems innovations.

Wise decisions at the presidential/congressional level require vigorous and informed debate. Choosing among complex weapons systems requires information on their functions and role in the nation's defense. The starting point must be, as stated by David Packard, Hewlett Packard chairman, a requirement that the Department of Defense "make a more rigorous effort to match strategy with resources. . . . For too long, the defense establishment has determined budgets without sufficient regard to long-term plans, and has developed defense plans without any regard to budgets."[18] *A critical part of the necessary long-term planning must be selection of which potential weapons systems innovations to develop.* Major systems both reflect and drive defense strategy or lack of strategy. Control of major weapons systems is only possible prior to congressional authorization of their development. As discussed in chapter 8, once a prime development contract has been let, a political system that is nearly impossible to control develops around the weapons system.

Recent efforts to deal with the problems faced by the nation's defense establishment have focused on two needs: a chairman of the Joint Chiefs of Staff who is independent of the services, and a procurement process that is independent of the services (i.e., a separate assistant secretary of defense for procurement). Both these foci reflect the feeling that weapons should be developed and deployed in response to the nation's security needs, not the needs of the services, which are presently reflected in the consensus decisions of the special-interest policy system for defense.

Put concisely, the continuous synthetic innovation of weapons systems during the post–World War II period has made the individual services obsolete. Modern weapons systems have made the boundaries among land, sea, and air obsolete, and thus the fundamental rationale for the services is inconsistent with the reality of present and future defense needs. Modern technology requires that defense strategy and the development of weapons systems be approached holistically.

The President and Congress must establish a consolidated national defense agency directly responsible to a single commander with the authority to control all defense functions. Hesitant efforts aimed at incrementally increasing the power of the chairman of the Joint Chiefs of Staff and the control of procurement by a single official will not work. Anything less than a consolidated national defense force mandated by the president and Congress will be gutted by the power and inertia of the existing special-interest policy system for defense. Reducing it to its essentials, the president and Congress must fundamentally restructure the special-interest policy system for defense. Only such a restructuring will allow for the establishment of the necessary priorities among the next generation of weapons systems, given finite and shrinking resources.

We have not investigated the special-interest policy systems for medicine and agriculture in the same detail as we have that for defense, in part because they are not so important in their impact on the resources necessary to achieve improved national competitiveness in the export race. In principle, however, in those sectors the same need exists for providing a strategic review of innovation options and for providing an informed basis for public debate on the choices. For example, the development of ever more sophisticated diagnostic and treatment techniques and technologies in medicine needs to be assessed in terms of the nation's and world's health-care priorities.

The ability to sustain the elderly in a condition that many of them find unattractive must be questioned. That development becomes an ever more serious problem when the resources required grow very large. The ability of other countries to provide health care at lower costs than the United States reflects, in part, the technological-fix orientation of the U.S. approach to health care. As with defense, the need is to ensure that choices are made in the context of a public debate emphasizing that the number of possible technological options is too large to be supported.

Confronting the Secular Trinity

America's contemporary dilemma is caused not by a lack of innovative capabilities but rather by a mismatch between our capabilities and those of a growing number of foreign competitors. If the experience of the 1980s has been grim, the future will be even more so unless a major restructuring and reallocation of our innovative capabilities occurs. The rising innovative capabilities of the NICs are likely both to increase the number of competitors and accelerate the rate of commercial innovation. Under present circumstances U.S. innovations in defense and medicine will generate ever-growing domestic demands for resources.

Reduced to the essentials, the United States has a simple choice: it must either consume less or produce more. While that simple choice is a real one over the long term, in the near term it is not so simple, for in the near term Americans must both consume less *and* produce more. That will be difficult and probably not politically possible, short of a severe economic disruption. Pushed to its fundamentals, the inability of the nation to reduce consumption, reallocate resources, and build a capability for export-oriented innovation has its roots in the secular trinity.

Recall from earlier chapters that successful societies have ideologies that are accepted by most of their members and that provide an accurate picture of reality. The secular trinity is widely subscribed to but provides a fundamentally flawed picture of contemporary reality, and each time we use it to address a problem, we make the problem worse. In the synthetic society success flows from the group, from public- and private-sector cooperation, and from innovation.

The centrality to the American belief system of the individual, the free market, and efficiency can hardly be exaggerated. It will not be easy for many Americans to accept the group as the source of creativity, to accept public- and private-sector cooperation as the vehicle to competitiveness, and to accept innovation as the instrument of success. This is because societies do not easily give up their ideologies. On the contrary, in the face of problems the instinctive response is to seek solutions in a more thoroughgoing application of existing ideology. The likelihood is high, then, that as the nation's competitiveness deteriorates, we will sink back into the old, tired, and fruitless debate between free enterprise and greater government control. In times of social and economic disruption and disorder, the tendency is for advocates of each view to seek answers in a purer application of their discredited prescriptions.

With the first compelling evidence of the nation's eroding economic position in the 1970s, Americans clearly sought answers in a purer application of the secular trinity. The economic problem was widely seen as flowing from too much government interference and control. Taxes were too high and inappropriately levied, health and environmental regulations were too burdensome, and welfare programs were rewarding people for not working. The answer was lower taxes, less regulation, less support for the lazy, and a willingness to require that the inefficient—whether organizations or individuals—suffer the consequences of their inefficiency. Later on, the conventional wisdom held that if tax cuts, deregulation, cuts in welfare, bankruptcies, and the loss of jobs didn't work the first time around, it was because we had not gone far enough. The answer, according to the secular trinity, was a second round of more significant initiatives.

As the United States remains uncompetitive in the world market, the pressure for just the opposite set of actions will likely grow. Given the nation's history and the pervasive power of the secular trinity, we will not frame the debate in socialist terms. The effect will be the same, however. In a piecemeal fashion proposals will grow for government to manage the economy, while those supporting such efforts will find ways to explain what is happening in terms of the secular trinity—for example, the support of semiconductor manufacturing by the DOD. The result will be a new evolution of the debate between advocates of two flawed and demonstrably unsuccessful approaches, both formulated to provide understanding and management of an industrial reality that no longer exists.

An accurate picture of the synthetic society is necessary if we are to meet the challenge we face in the last decade of this century. Lester Thurow has characterized the nature of the challenge as follows:

Needed Steps

Let no one think that altering two decades of low productivity growth is a trivial task. It is not going to be done with minor policy measures adopted in Washington. The changes that will be required can only be described as fundamental structural change. Every American and almost every American institution will have to be willing to change if Americans are to meet the economic challenges facing them as they prepare to enter the third millennium.[19]

The fundamental changes needed are three: (1) we must build an organizational complex capable of continuous synthetic innovation of products and processes aimed at winning in the international market; (2) we must reallocate resources from defense, medicine, and agriculture to synthetic innovation aimed at the commercial market; and (3) we must move beyond the constraints imposed by the secular trinity— that is, we must recognize that groups, public- and private-sector cooperation, and innovation are the essential conceptual reference points in the synthetic society.

What are the prospects of these changes occurring? Unfortunately, a search in history for examples of nations making such changes without a severe crisis leaves one wanting. The distressing evidence is that countries must lose wars, experience successful revolutions, or suffer severe economic disruptions (e.g., major depressions) before major ideological and policy changes occur and before major resource reallocations can be made.

This book has been written with the hope that it might make a small contribution to a process of rethinking and orderly change in American policy and ideas—a process that would make it possible for the United States to regain its competitiveness without the pain and disorder that will be associated with an economic disruption. In truth, one must be impressed with the numerous examples of flexibility and openness that have made it possible for some American companies and some subsectors to meet the challenge of the export race. Upon reflection, however, one sees that the number of such success stories is small, when measured against need. Rather than developing the capacity to out-innovate our commercial competitors, we have retained our focus on defense, medicine, and agriculture and added a process of continuous manipulation of corporate America's finances and ownership. Thus one must be pessimistic about America's ability to respond to the synthetic reality in the absence of crisis.

In that light, it is my hope that this book will make a small contribution to the understanding and debate necessary to develop an agenda for how to respond when the economic disruption occurs. If history has any message, it is that when social crises hit, political systems respond more successfully when they have clear, widely debated options to choose from.

APPENDIX

FIGURE 1

U.S. Trade in Synthetic Resins, Rubbers & Plastics, 1980–1985

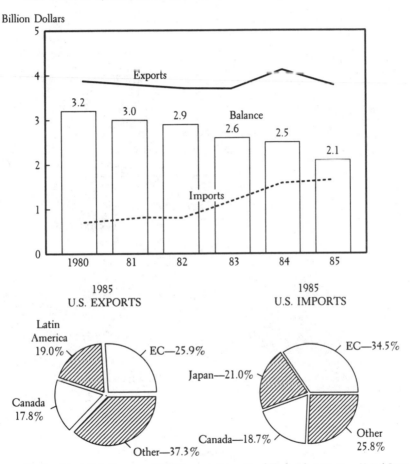

1985
U.S. EXPORTS

1985
U.S. IMPORTS

SOURCE: Reprinted from U.S. Department of Commerce, International Trade Administration, *United States Trade: Performance in 1985 and Outlook* (Washington, D.C.: Government Printing Office, 1986), 20.

FIGURE 2

U.S. Trade in Nonmetallic Mineral Manufactures, 1980–1985

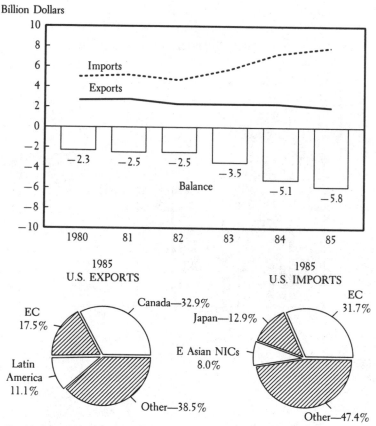

SOURCE: Reprinted from U.S. Department of Commerce, International Trade Administration, *United States Trade: Performance in 1985 and Outlook* (Washington, D.C.: Government Printing Office, 1986), 20.

Appendix

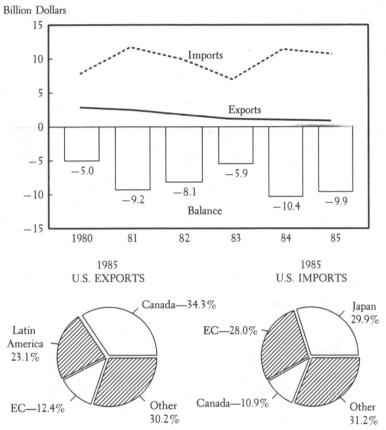

FIGURE 3

U.S. Trade in Iron & Steel Mill Products, 1980–1985

Billion Dollars

Imports

Exports

−5.0

−9.2

−8.1

−5.9

−10.4

−9.9

Balance

1980   81   82   83   84   85

1985
U.S. EXPORTS

1985
U.S. IMPORTS

Canada—34.3%

Latin
America
23.1%

EC—12.4%

Other
30.2%

Japan
29.9%

EC—28.0%

Canada—10.9%

Other
31.2%

source: Reprinted from U.S. Department of Commerce, International Trade Administration, *United States Trade: Performance in 1985 and Outlook* (Washington, D.C.: Government Printing Office, 1986), 20.

FIGURE 4
U.S. Trade in Nonferrous Metals, 1980–1985

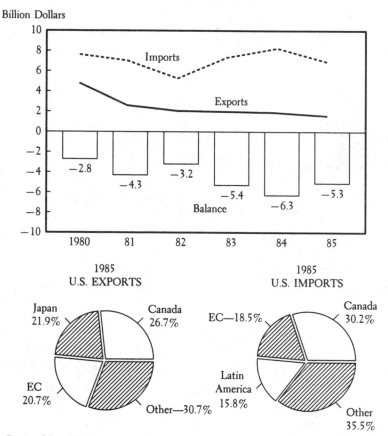

SOURCE: Reprinted from U.S. Department of Commerce, International Trade Administration, *United States Trade: Performance in 1985 and Outlook* (Washington, D.C.: Government Printing Office, 1986), 20.

Appendix

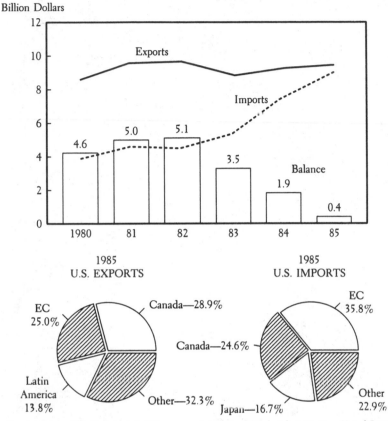

FIGURE 5

U.S. Trade in Power Generating Machinery, 1980–1985

Billion Dollars

Exports

Imports

4.6   5.0   5.1   3.5   Balance   1.9   0.4

1980   81   82   83   84   85

1985
U.S. EXPORTS

1985
U.S. IMPORTS

Canada—28.9%

EC
25.0%

Latin
America
13.8%

Other—32.3%

EC
35.8%

Canada—24.6%

Other
22.9%

Japan—16.7%

SOURCE: Reprinted from U.S. Department of Commerce, International Trade Administration, *United States Trade: Performance in 1985 and Outlook* (Washington, D.C.: Government Printing Office, 1986), 21.

FIGURE 6

U.S. Trade in Specialized Industrial Machinery, 1980–1985

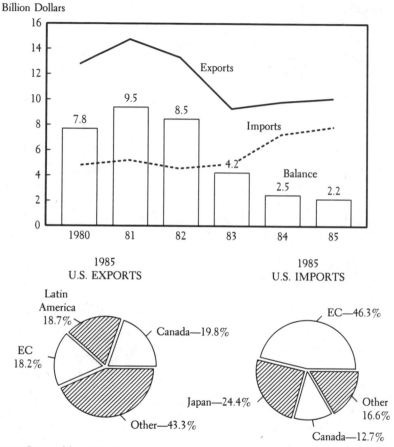

SOURCE: Reprinted from U.S. Department of Commerce, International Trade Administration, *United States Trade: Performance in 1985 and Outlook* (Washington, D.C.: Government Printing Office, 1986), 21.

Appendix

FIGURE 7

*U.S. Trade in Mechanical Equipment and
Non-Electrical Machinery Parts, 1980–1985*

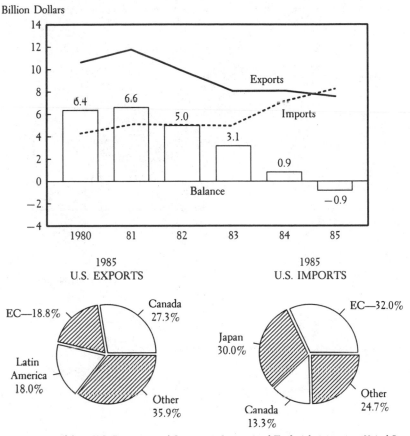

SOURCE: Reprinted from U.S. Department of Commerce, International Trade Administration, *United States Trade: Performance in 1985 and Outlook* (Washington, D.C.: Government Printing Office, 1986), 21.

FIGURE 8

U.S. Trade in Office & ADP Machines, 1980–1985

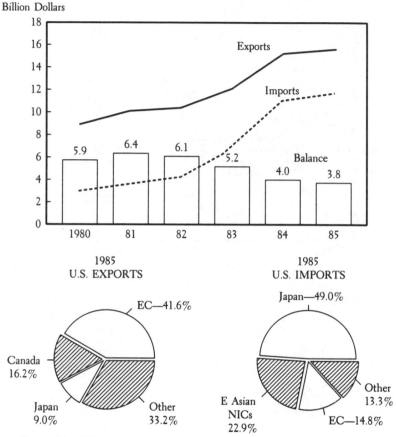

SOURCE: Reprinted from U.S. Department of Commerce, International Trade Administration, *United States Trade: Performance in 1985 and Outlook* (Washington, D.C.: Government Printing Office, 1986), 21.

Appendix

FIGURE 9
*U.S. Trade in Telecommunications & Sound
Reproducing Equipment, 1980–1985*

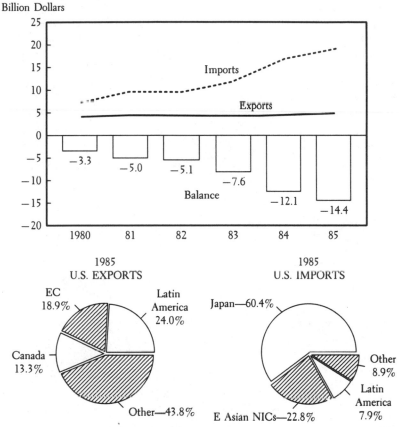

source: Reprinted from U.S. Department of Commerce, International Trade Administration, *United States Trade: Performance in 1985 and Outlook* (Washington, D.C.: Government Printing Office, 1986), 22.

FIGURE 10
*U.S. Trade in Integrated Circuits & Other
Electrical Equipment, 1980–1985*

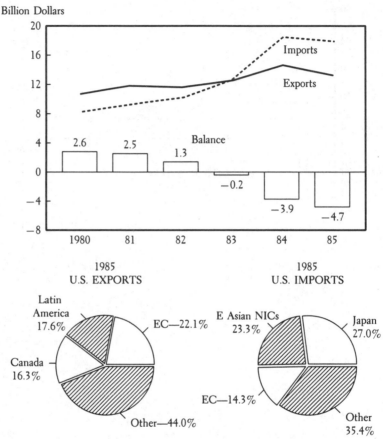

SOURCE: Reprinted from U.S. Department of Commerce, International Trade Administration, *United States Trade: Performance in 1985 and Outlook* (Washington, D.C.: Government Printing Office, 1986), 22.

Appendix

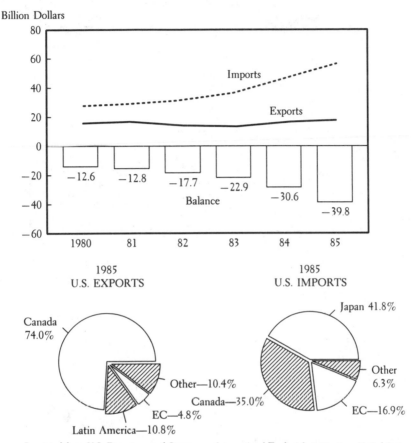

FIGURE 11

U.S. Trade in Motor Vehicles & Parts, 1980–1985

Billion Dollars

Imports

Exports

−12.6   −12.8   −17.7   −22.9   −30.6

Balance

−39.8

1980   81   82   83   84   85

1985
U.S. EXPORTS

1985
U.S. IMPORTS

Canada
74.0%

Other—10.4%

Canada—35.0%

EC—4.8%

Latin America—10.8%

Japan 41.8%

Other
6.3%

EC—16.9%

SOURCE: Reprinted from U.S. Department of Commerce, International Trade Administration, *United States Trade: Performance in 1985 and Outlook* (Washington, D.C.: Government Printing Office, 1986), 22.

243

FIGURE 12

U.S. Trade in Aircraft & Other Transport Equipment, 1980–1985

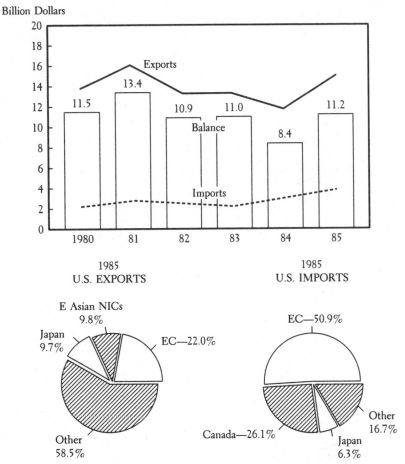

SOURCE: Reprinted from U.S. Department of Commerce, International Trade Administration, *United States Trade: Performance in 1985 and Outlook* (Washington, D.C.: Government Printing Office, 1986), 22.

Appendix

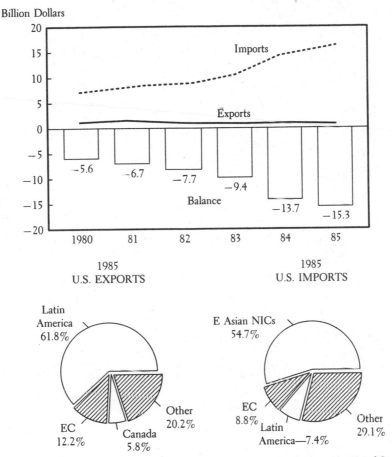

FIGURE 13

U.S. Trade in Apparel & Accessories, 1980–1985

SOURCE: Reprinted from U.S. Department of Commerce, International Trade Administration, *United States Trade: Performance in 1985 and Outlook* (Washington, D.C.: Government Printing Office, 1986), 23.

245

FIGURE 14
U.S. Trade in Footwear, 1980–1985

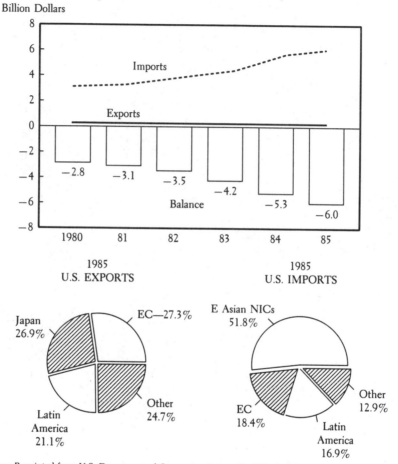

Billion Dollars

| | 1980 | 81 | 82 | 83 | 84 | 85 |

Imports

Exports

−2.8   −3.1   −3.5   −4.2   −5.3   −6.0

Balance

1985
U.S. EXPORTS

Japan 26.9%   EC—27.3%   Latin America 21.1%   Other 24.7%

1985
U.S. IMPORTS

E Asian NICs 51.8%   EC 18.4%   Latin America 16.9%   Other 12.9%

SOURCE: Reprinted from U.S. Department of Commerce, International Trade Administration, *United States Trade: Performance in 1985 and Outlook* (Washington, D.C.: Government Printing Office, 1986), 23.

Appendix

FIGURE 15
U.S. Trade in Professional, Scientific & Control Instruments, 1980–1985

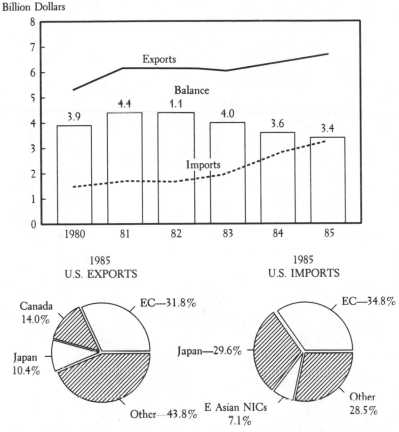

Billion Dollars

1985
U.S. EXPORTS

1985
U.S. IMPORTS

SOURCE: Reprinted from U.S. Department of Commerce, International Trade Administration, *United States Trade: Performance in 1985 and Outlook* (Washington, D.C.: Government Printing Office, 1986), 23.

FIGURE 16

U.S. Trade in Miscellaneous Consumer Manufactures, 1980–1985

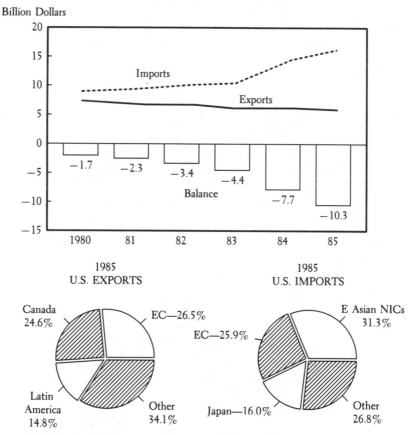

SOURCE: Reprinted from U.S. Department of Commerce, International Trade Administration, *United States Trade: Performance in 1985 and Outlook* (Washington, D.C.: Government Printing Office, 1986), 23.

NOTES

CHAPTER 1

1. U.S. National Commission on Excellence in Education, *A Nation at Risk: The Imperative for Educational Reform: A Report to the Nation and the Secretary of Education* (Washington, D.C.: Department of Education, 1983).

2. James C. Abegglen and George Stalk, Jr., *Kaisha: The Japanese Corporation* (New York: Basic Books, 1985), 134.

3. Don E. Kash and Robert W. Rycroft, *U.S. Energy Policy: Crisis and Complacency* (Norman: University of Oklahoma Press, 1984), 197–98.

4. "Two Key Senators Join in Assault on the Military," *New York Times*, 6 October 1985, sec. 1.

5. "Goldwater-Nichols Department of Defense Reorganization Act of 1986," *U.S. Code Congressional and Administrative News* 1 (1986): 992–1075.

6. Michael Harrington, *The Accidental Century* (Baltimore: Penguin Books, 1965), 16.

7. Robert B. Reich, *The Next American Frontier* (New York: Times Books, 1983), 8.

8. Rachel Carson, *Silent Spring* (Greenwich, CN: Fawcett Publications, 1962).

9. "What Robins Has Wrought," *New York Times*, 13 December 1987, sec. 3.

10. U.S. Congress, Office of Technology Assessment, *Technologies and Management Strategies for Hazardous Waste Control* (Washington, D.C.: Government Printing Office, 1983), 6.

11. Lester Lave and Arthur Upton, eds., *Toxic Chemicals, Health and the Environment* (Baltimore: Johns Hopkins University Press, 1987).

12. American Association for the Advancement of Science, *AAAS Report XIII: Research and Development FY 1989* (Washington, D.C.: AAAS, 1988), 7.

13. U.S. Bureau of the Census, *Statistical Abstract of the United States: 1987* (Washington, D.C.: Government Printing Office, 1986), 84, 319, 416, 636.

14. "Competitiveness: 23 Leaders Speak Out," *Harvard Business Review*, July–August 1987, 107–8.

15. U.S. Department of Commerce, International Trade Administration, *United States Trade: Performance in 1985 and Outlook* (Washington, D.C.: Government Printing Office, 1986), 111. Data for 1987 provided by U.S. Department of Commerce, International Trade Administration.

16. U.S. Department of Commerce, International Trade Administration, *United States Trade: Performance in 1985*, 111. Data for 1988 provided by U.S. Department of Commerce, International Trade Administration.

17. Peter G. Peterson, "The Morning After," *Atlantic Monthly* 260 (October 1987): 50.

18. Richard M. Cyert and David C. Mowery, eds., *Technology and Employment: Innovation and Growth in the U.S. Economy* (Washington, D.C.: National Academy Press, 1987), 79.

19. Edward O'Toole, "The Globalization of the Industrialized Economies," *Barrons* 67 (4 May 1987): 45–56.

20. U.S. Department of Commerce, International Trade Administration, *United States Trade: Performance in 1985*, 15.

21. Ibid.

22. Data provided by U.S. Department of Commerce, International Trade Administration.

23. Ibid.

24. Harold A. Schneiderman, "Innovation in Agriculture," *The Bridge* 17 (Spring 1987): 2.

25. *Japan's Industrial Policies* (Washington, D.C.: Japan Economic Institute, 1984), 7.

26. Bruce R. Scott, "U.S. Competitiveness: Concepts, Performance and Implications," in *U.S. Competitiveness in the World Economy*, ed. Bruce R. Scott and George C. Lodge (Cambridge, MA: Harvard Business School Press, 1985), 47–49.

CHAPTER 2

1. John Kenneth Galbraith, *The New Industrial State* (Boston: Houghton Mifflin, 1967), 19.
2. Robert B. Reich, *The Next American Frontier* (New York: Times Books, 1983), 22.
3. Barry Commoner, *The Closing Circle: Nature, Man, and Technology* (New York: Alfred A. Knopf, 1971), 129–32.
4. J. J. Beer, *The Emergence of the German Dye Industry* (Urbana: University of Illinois Press, 1959).
5. Commoner, *Closing Circle*, 132.
6. "Science and Technology," *Economist*, 7 September 1985, 95–96.
7. Christopher Hill, "Technological Innovation: Agent of Growth and Change," in *Technological Innovation for a Dynamic Economy*, ed. Christopher Hill and James M. Utterback (New York: Pergamon Press, 1979), 3.
8. Will Crutchfield, "Next Home Stereo Advance: Digital Tape Cassettes in 1987," *New York Times*, 24 October 1986, sec. 1.
9. "Sony's Challenge," *Business Week*, 1 June 1987, 69.
10. James C. Abegglen and George Stalk, Jr., *Kaisha: The Japanese Corporation* (New York: Basic Books, 1985).
11. Fred Kaplan, *The Wizards of Armageddon* (New York: Simon and Schuster, 1983), 299–301.
12. Ibid., 10.
13. Samuel F. Wells, Jr., "America and the 'MAD' World," *The Wilson Quarterly* 1 (Autumn 1977): 64–65.
14. Vernon Van Dyke, *Pride and Power: The Rationale of the Space Program* (Urbana: University of Illinois Press, 1964), 104.
15. Harvey Brooks, "Technology as a Factor in U.S. Competitiveness," in *U.S. Competitiveness in the World Economy*, ed. Bruce R. Scott and George C. Lodge (Cambridge, MA: Harvard Business School Press, 1985), 330–34.
16. *Japan's Industrial Policies* (Washington, D.C.: Japan Economic Institute, 1984).
17. Fred V. Guterl, "Star Wars Is Bad for Business," *Dun's Business Month* 128 (September 1986): 56.
18. Paul Lewis, "Military Spending Questioned," *New York Times*, 11 November 1986, sec. 4.
19. Ibid.
20. U.S. Congress, Office of Technology Assessment, *Development and Diffusion of Commercial Technologies: Should the Federal Government Redefine Its Role?* (Washington, D.C.: Government Printing Office, 1984), 21.
21. Ibid., 19.

CHAPTER 3

1. A. Fredga, "Presentation Speech by A. Fredga, Member of the Nobel Committee for Chemistry of the Royal Swedish Academy of Sciences," *Nobel Lectures: Chemistry, 1965* (Amsterdam: Elsevier, 1972), 97.
2. Robert H. Waterman, Jr., *The Renewal Factor: How the Best Get and Keep the Competitive Edge* (New York: Bantam Books, 1987), 41–42.
3. Derek J. de Solla Price, *Little Science, Big Science* (New York: Columbia University Press, 1963), 1–19.
4. John Kenneth Galbraith, *The New Industrial State*, 3rd ed. (Boston: Houghton Mifflin, 1978), 64.

Notes

5. Ibid., 62.

6. David Halberstam, *The Reckoning* (New York: Morrow, 1986), 499.

7. Waterman, Jr., *Renewal Factor*, 157–58.

8. Roland W. Schmitt, "The Japanese Style" (Photocopy of typescript, Rensselaer Polytechnic Institute, 1987), 4–6.

9. U.S. Congress, Office of Technology Assessment, *Development and Diffusion of Commercial Technologies: Should the Federal Government Redefine Its Role?* (Washington, D.C.: Government Printing Office, 1984), 3–4.

10. Waterman, *Renewal Factor*, 43.

11. Robert B. Reich, *Tales of a New America* (New York: Times Books, 1987), 124.

12. Akio Morita, *Made in Japan* (London: Fontana/Collins, 1987), 115.

13. Reich, *Tales of a New America*, 88.

14. Deborah Shapley and Rustum Roy, *Lost at the Frontier: U.S. Science and Technology Policy Adrift* (Philadelphia: ISI Press, 1985), 33.

15. Halberstam, *Reckoning*, 693.

16. Ibid., 693–94.

CHAPTER 4

1. Lester C. Thurow, *The Zero-Sum Solution: Building A World Class American Economy* (New York: Simon and Schuster, 1985), 125.

2. Akio Morita, *Made in Japan* (London: Fontana/Collins, 1987), 150.

3. Robert B. Reich, *Tales of a New America* (New York: Times Books, 1987), 123.

4. Robert H. Waterman, Jr., *The Renewal Factor: How the Best Get and Keep the Competitive Edge* (New York: Bantam Books, 1987), 179–80.

5. John Kenneth Galbraith, *The New Industrial State* (Boston: Houghton Mifflin, 1967), 62.

6. Reich, *Tales of a New America*, 222–32.

7. George C. Lodge and William C. Crum, "The Pursuit of Remedies," in *U.S. Competitiveness in the World Economy*, ed. Bruce R. Scott and George C. Lodge (Cambridge, MA: Harvard Business School Press, 1985), 495.

8. Thurow, *Zero-Sum Solution*, 90–109.

9. U.S. Department of Commerce, International Trade Administration, *United States Trade: Performance in 1984 and Outlook* (Washington, D.C.: Government Printing Office, 1985), 13.

10. Data for 1987 provided by U.S. Department of Commerce, International Trade Administration.

11. Harvey Brooks, "Technology as a Factor in U.S. Competitiveness," in *U.S. Competitiveness in the World Economy*, ed. Bruce R. Scott and George C. Lodge (Cambridge, MA: Harvard Business School Press, 1985), 335–36.

CHAPTER 5

This chapter builds on the work of Hugh Heclo, "Issue Networks and the Executive Establishment," in *The New American Political System*, ed. Anthony King (Washington, D.C.: American Enterprise Institute, 1978), 87–124; Randall B. Ripley and Grace A. Franklin, *Congress, the Bureaucracy and Public Policy*, 3rd ed. (Homewood, IL: Dorsey Press, 1984); and Don E. Kash and Robert W. Rycroft, *U.S. Energy Policy: Crisis and Complacency* (Norman: University of Oklahoma Press, 1984), 21–45.

1. Don K. Price, *Government and Science: Their Dynamic Relation in American Democracy* (New York: Oxford University Press, 1962), 27–31.

2. Alvin M. Weinberg, "Can Technology Replace Social Engineering?" in *Technology and Man's Future*, 3rd ed., ed. Albert H. Teich (New York: St. Martin's Press, 1981), 29–39.

3. Vernon Van Dyke, *Pride and Power* (Urbana: University of Illinois Press, 1964).

4. Kash and Rycroft, *U.S. Energy Policy*, 46–77.

5. Price, *Government and Science*, 10–11.

6. Dorothy Nelkin, "Science and Technology and Political Conflict: Analyzing the Issues," in

Controversy: Politics of Technical Decisions, 2nd ed., ed. Dorothy Nelkin (Beverly Hills, CA: Sage, 1984), 12.

7. John W. Kingdom, *Agendas, Alternatives and Public Policies* (Boston: Little, Brown, 1984), 209.

8. Robert W. Rycroft and Joseph S. Szyliowicz, "The Technological Dimension of Decision-Making: The Case of the Aswan High Dam," *World Politics* 33 (October 1980): 36–61.

9. Thomas S. Kuhn, *The Structure of Scientific Revolutions* (Chicago: University of Chicago Press, 1962), 1–22.

10. James Q. Wilson, "The Rise of the Bureaucratic State," *Public Interest* 41 (Fall 1975): 77–103.

11. Hans H. Landsberg, "Let's All Play Energy Policy!" *Daedalus* 109 (Summer 1980): 84.

12. Fred V. Guterl, "Star Wars Is Bad for Business," *Dun's Business Month* 128 (September 1986): 56–58.

13. David Packard, "Improving Weapons Acquisitions: What the Defense Department Can Learn from the Private Sector," *Policy Review* 37 (Summer 1986): 11–15.

14. Paul Kennedy, *The Rise and Fall of the Great Powers* (New York: Random House, 1987), 347–535.

CHAPTER 6

1. James Phinney Baxter, *Scientists Against Time* (Boston: Little, Brown, 1947), vii–viii. (Foreword by Vannevar Bush)

2. Ibid., 145.

3. Daniel S. Greenberg, *The Politics of Pure Science* (New York: New American Library, 1967), 90–91.

4. Baxter, *Scientists Against Time*.

5. Greenberg, *Politics of Pure Science*, 81.

6. Harvey Brooks, *The Government of Science* (Cambridge, MA: MIT Press, 1968), 21.

7. Ibid., 24.

8. Ibid., 20–21.

9. Ibid., 20.

10. H. L. Nieburg, *In the Name of Science* (New York: Quadrangle Books, 1966), 218–19.

11. Ibid.

12. Greenberg, *Politics of Pure Science*, 51–67.

13. Ibid., 26–37.

14. Brooks, *Government of Science*, 21.

15. Don K. Price, *The Scientific Estate* (Boston: Belknap Press of Harvard University Press, 1967), 68–81.

16. Richard Rhodes, *The Making of the Atomic Bomb* (New York: Simon and Schuster, 1987).

17. William J. Broad, review of *The Making of the Atomic Bomb*, by Richard Rhodes, *New York Times Book Review*, 8 February 1987.

18. Baxter, *Scientists Against Time*.

19. Brooks, *Government of Science*, 22.

20. Greenberg, *Politics of Pure Science*, 70–71.

21. Brooks, *Government of Science*, 22.

22. Greenberg, *Politics of Pure Science*, 70–71.

23. Richard Bolt, personal interview, 17 February 1987.

24. Greenberg, *Politics of Pure Science*, 68–80.

25. Ibid., 78.

26. Baxter, *Scientists Against Time*, 241–42.

27. Ibid., 247–48.

28. Ibid., 249–50.

29. Arthur H. Compton, *Atomic Quest* (New York: Oxford University Press, 1956), 94–95.

30. Clarence H. Danhof, *Government Contracting and Technological Change* (Washington, D.C.: Brookings Institution, 1968), 75.

31. Brooks, *Government of Science*, 24.

Notes

32. *Science and Technology Data Book, 1988* (Washington, D.C.: National Science Foundation, Division of Science Resources Studies, 1988), 15.

33. Nieburg, *In the Name of Science,* 185–86.

CHAPTER 7

1. Testimony before U.S. Congress, House Committee on Science and Astronautics, Subcommittee on Science, Research, and Development, *Toward a Science Policy for the United States* (Washington, D.C.: Government Printing Office, 1970). Cited in W. Henry Lambright, *Governing Science and Technology* (New York: Oxford University Press, 1976), 15–16.

2. U.S. Congress, House Committee on Science and Technology, *Science Support by the Department of Defense* (Washington, D.C.: Government Printing Office, 1986), 360–61.

3. Don K. Price, *Government and Science: Their Dynamic Relation in American Democracy* (New York: Oxford University Press, 1962), 65–94.

4. American Association for the Advancement of Science, *AAAS Report XIII: Research and Development FY 1989* (Washington, D.C.: AAAS, 1988), 8.

5. Clarence H. Danhof, *Government Contracting and Technological Change* (Washington, D.C.: Brookings Institution, 1968), 17–18.

6. Ibid., 31.

7. Ibid., 131–84.

8. Harold Orlans, *Contracting for Atoms* (Washington, D.C.: Brookings Institution, 1967), 128.

9. American Association for the Advancement of Science, *AAAS Report XIII,* 7.

10. Ibid., 38.

11. Ibid., 132.

12. William J. Broad, "Weapons in Space/The Origins of 'Star Wars': Reagan's 'Star Wars' Bid: Many Ideas Converging," *New York Times,* 4 March 1985, sec. 1.

13. C. A. Zraket, "Uncertainties in Building a Strategic Defense Initiative," *Science* 235 (27 March 1987): 1600.

CHAPTER 8

1. These contract types roughly parallel five types of relationships between government and private institutions identified in Don K. Price, *Government and Science: Their Dynamic Relation in American Democracy* (New York: Oxford University Press, 1962), 68–75.

2. *Science and Technology Data Book, 1988* (Washington, D.C.: National Science Foundation, Division of Science Resources Studies, 1988), 15.

3. American Association for the Advancement of Science, *AAAS Report IX: Research and Development FY 1985* (Washington, D.C.: AAAS, 1984), 36, 42.

4. American Association for the Advancement of Science, *AAAS Report XIII: Research and Development FY 1989* (Washington, D.C.: AAAS, 1988), 6.

5. Warren O. Hagstrom, *The Scientific Community* (New York: Basic Books, 1965), 12–42.

6. H. L. Nieburg, *In the Name of Science* (New York: Quadrangle Books, 1966), 200–17.

7. Clarence H. Danhof, *Government Contracting and Technological Change* (Washington, D.C.: Brookings Institution, 1968), 110.

8. Bruce L. R. Smith, *The Rand Corporation: Case Study of a Nonprofit Advisory Corporation* (Cambridge, MA: Harvard University Press, 1966).

9. Fred Kaplan, *The Wizards of Armageddon* (New York: Simon and Schuster, 1983), 86–88.

10. Harold Orlans, *Contracting for Atoms* (Washington, D.C.: Brookings Institution, 1967).

11. Danhof, *Government Contracting and Technological Change,* 374–79.

12. U.S. Congress, Joint Committee on Atomic Energy, *Investigations into the United States Atomic Energy Project,* 81st Cong., 1st sess., Pt. 20, pp. 803, 827; and David Lilienthal, *Change, Hope, and the Bomb* (Princeton: Princeton University Press, 1963), 79. Cited in Orlans, *Contracting for Atoms,* 17.

13. Orlans, *Contracting for Atoms*, 15.

14. Price, *Government and Science*, 68–69.

15. David Packard, "Improving Weapons Acquisition: What the Defense Department Can Learn from the Private Sector," *Policy Review* 37 (Summer 1986): 11–15.

16. Ibid., 12–13.

17. Mel Horwitch, *Clipped Wings: The American SST Conflict* (Cambridge, MA: MIT Press, 1982).

18. Nieburg, *In the Name of Science*, 41–49.

CHAPTER 9

1. H. L. Nieburg, *In the Name of Science* (New York: Quadrangle Books, 1966), 214–43.

2. Merton J. Peck and Frederick M. Scherer, *The Weapons Acquisition Process: An Economic Analysis* (Cambridge, MA: Graduate School of Business Administration, Harvard University, 1962), 24.

3. Clarence H. Danhof, *Government Contracting and Technological Change* (Washington, D.C.: Brookings Institution, 1968); and J. Ronald Fox, *Arming America: How the U.S. Buys Weapons* (Cambridge, MA: Harvard University Press, 1974).

4. Robert J. Art, *The TFX Decision: McNamara and the Military* (Boston: Little, Brown, 1968), 91.

5. Art, *TFX Decision*.

6. Don Dwiggins, *The SST: Here it Comes, Ready or Not* (Garden City, NY: Doubleday, 1968), 173–74.

7. Art, *TFX Decision*, 132.

8. Berkeley Rice, *The C-5A Scandal: An Inside Story of the Military-Industrial Complex* (Boston: Houghton Mifflin, 1971), 178–213.

9. Charles H. Ferguson, "Obsolete Arms Production, Obsolete Military," *New York Times*, 11 April 1988.

10. David Packard, "Improving Weapons Acquisition: What the Defense Department Can Learn from the Private Sector," *Policy Review* 37 (Summer 1986): 15.

CHAPTER 10

1. Don K. Price, *Government and Science: Their Dynamic Relation in American Democracy* (New York: Oxford University Press, 1962), 10.

2. Dennis J. Palumbo, *Public Policy in America: Government in Action* (San Diego: Harcourt Brace Jovanovich, 1988), 60–206.

3. Victoria A. Harden, *Inventing the NIH: Federal Biomedical Research Policy* (Baltimore: Johns Hopkins University Press, 1986), 1–5.

4. American Association for the Advancement of Science, *AAAS Report XII: Research and Development FY 1988* (Washington, D.C.: AAAS, 1987), 35.

5. Harden, *Inventing the NIH*, 1.

6. Burton A. Weisbrod, *Economics and Medical Research* (Washington, D.C.: American Enterprise Institute for Public Policy Research, 1983), 31–42.

7. Robert Pear, "Hospitals' Medicare Profits Drop," *New York Times*, 28 January 1988.

8. Price, *Government and Science*, 12–13.

9. Ross B. Talbot and Don F. Hadwiger, *The Policy Process in American Agriculture* (San Francisco: Chandler Publishing, 1968), 31–32.

10. U.S. Bureau of the Census, *Statistical Abstract of the United States: 1987* (Washington, D.C.: Government Printing Office, 1986), 418, 636.

11. Don F. Hadwiger, *The Politics of Agricultural Research* (Lincoln: University of Nebraska Press, 1982), 12–30.

12. American Association for the Advancement of Science, *AAAS Report XIII: Research and Development FY 1989* (Washington, D.C.: AAAS, 1988), 7.

Notes

13. Price, *Government and Science*, 14–15.

14. U.S. Congress, Office of Technology Assessment, *An Assessment of the United States Food and Agricultural Research System* (Washington, D.C.: Government Printing Office, 1983), 29–49.

15. Hadwiger, *Politics of Agricultural Research*, 21.

16. *Farm Price and Income Support Programs: Background Information*, 83–2 ENR prepared by Penelope C. Cate, (Washington, D.C.: U.S. Congressional Research Service, 1983), 36; *Farm Programs: An Overview of Price and Income Support, and Storage Programs*, GAO/RCED–88–84BR (Washington, D.C.: U.S. General Accounting Office, 1988), 58.

17. Barbara J. Culliton, "The (Private) University of NIH?" *Science* 239 (18 March 1988): 1364–65.

CHAPTER 11

1. Daniel I. Okimoto, "Regime Characteristics of Japanese Industrial Policy," in *Japan's High Technology Industries: Lessons and Limitations of Industrial Policy*, ed. Hugh Patrick (Seattle: University of Washington Press, 1986), 35.

2. Ibid.

3. Ken-ichi Imai, "Japan's Industrial Policy for High Technology Industry," in Patrick, *Japan's High Technology Industries*, 138.

4. Chalmers Johnson, *MITI and the Japanese Miracle: The Growth of Industrial Policy, 1925–1975* (Stanford, CA: Stanford University Press, 1982), 3–34.

5. Ibid., 311–12.

6. James C. Abegglen and George Stalk, Jr., *Kaisha: The Japanese Corporation* (New York: Basic Books, 1985), 41.

7. *Japan's Industrial Policies* (Washington, D.C.: Japan Economic Institute, 1984), 7.

8. Okimoto, "Regime Characteristics of Japanese Industrial Policy," 50.

9. Johnson, *MITI and the Japanese Miracle*, 199.

10. Leonard Lynn, "Japanese Research and Technology Policy," *Science* 233 (18 July 1986): 299.

11. Ibid.

12. Imai, "Japan's Industrial Policy for High Technology Industry," 152.

13. Akio Morita, *Made in Japan* (London: Fontana/Collins, 1987), 63–69.

14. Ibid., 64–73.

15. Ibid., 82.

16. Roland W. Schmitt, "The Japanese Style" (Photocopy of typescript, Rensselaer Polytechnic Institute, 1987), 1–2.

17. Abegglen and Stalk, *Kaisha*, 42–54.

18. David Halberstam, *The Reckoning* (New York: Morrow, 1986), 311–18.

19. Okimoto, "Regime Characteristics of Japanese Industrial Policy," 63.

20. Lester C. Thurow, *The Zero-Sum Solution: Building A World Class American Economy* (New York: Simon and Schuster, 1985), 54.

21. Abegglen and Stalk, *Kaisha*, 135.

22. Ibid.

23. Hugh Patrick, "Japanese High Technology Industrial Policy in Comparative Context," in Patrick, *Japan's High Technology Industries*, 14–18.

24. E. M. Krishna and C. F. Rao, "Is U.S. High Technology High Enough?" *Columbia Journal of World Business* 21 (Summer 1986): 51.

25. Abegglen and Stalk, *Kaisha*, 125.

26. Ibid., 120.

27. National Science Foundation, *The Science and Technology Resources of Japan: A Comparison with the United States* (Washington, D.C.: National Science Foundation, 1988), 48.

28. Lynn, "Japanese Research and Technology Policy," 298.

29. Gary R. Saxonhouse, "Industrial Policy and Factor Markets: Biotechnology in Japan and the United States," in Patrick, *Japan's High Technology Industries*, 126.

30. Lynn, "Japanese Research and Technology Policy," 298–99.

31. Ibid., 298.

32. Leonard L. Lederman, "Science and Technology Policies and Priorities: A Comparative Analysis," *Science* 237 (4 September 1987): 1127.

33. Lynn, "Japanese Research and Technology Policy," 298.

34. Marjorie Sun, "Japan's Inscrutable Research Budget," *Science* 238 (2 October 1987): 22.

35. Lynn, "Japanese Research and Technology Policy," 299.

36. Okimoto, "Regime Characteristics of Japanese Industrial Policy," 51.

37. Ibid., 63.

38. Fumio Kodama, "Technological Diversification of Japanese Industry," *Science* 233 (18 July 1986): 291–96.

39. Abegglen and Stalk, *Kaisha,* 232–35.

40. Okimoto, "Regime Characteristics of Japanese Industrial Policy," 92.

41. Robert B. Reich, *Tales of a New America* (New York: Times Books, 1987), 87.

42. Bruce R. Scott, "U.S. Competitiveness: Concepts, Performance, and Implications," in *U.S. Competitiveness in the World Economy,* ed. Bruce R. Scott and George C. Lodge (Cambridge, MA: Harvard Business School Press, 1985), 68–69. See also Bruce R. Scott, "National Strategies: Key to International Competition," in Scott and Lodge, *U.S. Competitiveness in the World Economy,* 91–93.

CHAPTER 12

1. Lester C. Thurow, *The Zero-Sum Solution: Building A World Class Economy* (New York: Simon and Schuster, 1985), 35–44.

2. Data provided by U.S. Department of Commerce, International Trade Administration.

3. Ibid.

4. Bruce R. Scott, "U.S. Competitiveness: Concepts, Performance, and Implications," in *U.S. Competitiveness in the World Economy,* ed. Bruce R. Scott and George C. Lodge (Cambridge, MA: Harvard Business School Press, 1985), 17.

5. Peter G. Peterson, "The Morning After," *Atlantic Monthly* 260 (October 1987): 50–52.

6. Roland W. Schmitt, "The Japanese Style" (Photocopy of typescript, Rensselaer Polytechnic Institute, 1987), 4–6.

7. Ibid., 4.

8. James C. Abegglen and George Stalk, Jr., *Kaisha: The Japanese Corporation* (New York: Basic Books, 1985), 112–14.

9. "R&D Policy After Reagan: First of a Special SGR Series," *Science and Government Report* 18 (15 January 1988): 1.

10. Don E. Kash, *Competitiveness, Technology and ORNL* (Oak Ridge, TN: Oak Ridge National Laboratory, 1987), 22.

11. Ibid., 21–29.

12. "Stevenson-Wydler Technology Innovation Act of 1980," *U.S. Code Congressional and Administrative News* 2 (1980): 2311–20.

13. Charles B. Watkins and Joan Wills, "State Initiatives to Encourage Economic Development through Technological Innovation," in *Technological Innovation Strategies for a New Partnership,* ed. Denis O. Gray, Trudy Solomon, and William Hetzner (Amsterdam: North-Holland, 1986), 69–87.

14. Jurgen Schmandt and Robert Wilson, *Promoting High Technology Industry: Initiatives and Policies for State Governments* (Boulder, CO: Westview Press, 1987).

15. "National Environmental Policy Act," *United States Statutes at Large* 83 Stat., 852–56, sec. 102 (C) (ii).

16. "Atomic Energy Act: Amendment," *United States Statutes at Large* 71 Stat., 576–79.

17. Thurow, *Zero-Sum Solution,* 35.

18. David Packard, "Improving Weapons Acquisition: What the Defense Department Can Learn from the Private Sector," *Policy Review* 37 (Summer 1986): 12.

19. Thurow, *Zero-Sum Solution,* 381–82.

INDEX

Index

Index

Index

Index

Index

Uncertainty of innovation, 39
Unemployment, 3
Union of Soviet Socialist Republics (USSR): arms race innovation and, 29, 31, 32, 126, 127, 150; Cuban missile crisis and, 29; policy-making system response to, 85; Strategic Defense Initiative (SDI) and, 10, 56
Universities: contract federal system with, 126; government and industry cooperation with, 21, 22, 104–108; master contracts in, 141–142; medical research and, 171; post-World War II research and, 122, 123; pre-World War II support for, 111–112; R&D base in, 221–222; research grants in, 139–141; role of scientists in, 114–116; spinoffs and, 33; special-interest policy and, 93, 94, 217, 218; strategic innovation and, 49, 50; World War II transition in, 112–120
University of California, 147
University of Chicago, 147
University of Tsukuba, 201
University Research Initiatives Program, 142

Videocassette recorder (VCR), 24, 195–196
Very large scale integrated circuits (VLSI), 185, 203

Walkman, 27, 193
War Department, 119–120, 132

Washington, George, 173
Waterman, Robert, 39, 47, 52, 67–68
Weapons systems: arms race innovation in, 30–33, 126; Congressional role in, 227; continuous synthetic innovation in, 126–128; procurement of, 5; R&D contracts for, 135–136; spinoffs of, 36–37; World War II development of, 79–80
Weinberg, Alvin, 84
Woodward, Robert, 39
Woolridge, Dean, 143
World War II: cooperation among government, industry, and universities and, 104–108, 123–124; defense policy and, 6, 79–80, 87, 90–91, 122–124; development of of synthetic capabilities in, 17–18, 79; Japanese manufacturing after, 188, 189–190; organizational networks before, 108–112; presidential/congressional policy making and, 87; R&D contract evolution after, 132–133; research efforts before, 108–112; special-interest policy making and, 90–91; transition in research during, 112–120; synthetic innovation and, 80, 104
Wozniak, Steven, 62
Wright-Patterson, 50

Xenophobia, 98, 99

Zraket, C. A., 137